FOR ALL PRACTICAL PURPOSES

STUDY GUIDE

Fifth Edition

Guided Readings prepared by
DAN REICH
Temple University

Practice Quizzes prepared by
JOHN EMERT and KAY ROEBUCK
Ball State University

W. H. Freeman and Company
New York

ACQUISITIONS EDITOR: Patrick Shriner
SUPPLEMENTS COORDINATOR: Vivien Weiss
PROJECT EDITOR: Jodi Isman
COMPOSITION: Passumpsic Publishing

ISBN: 0–7167–3699–3

Printed in the United States of America

First printing 1999

CONTENTS

PREFACE

FEATURES

Chapter Objectives

The Chapter Objectives feature is a list of skills that the student should master. Each objective is preceded by a toggle box that allows a student to "check off" a skill once he or she feels it has been mastered.

Guided Reading

Guided Reading covers each section of each chapter and is intended as a walk-through tutorial on each section. Students should read the corresponding chapter section in the textbook before working through the tutorial. Each Guided Reading section consists of a tutorial on the basics of that section's content. The section topics follow the order of content presentation in the textbook.

There are four types of Guided Reading sections: Key Idea, Question, Answer, and Explanation.

Key Ideas summarize crucial information from the section's content. Students should make sure they understand each of these points before proceeding to the Question sections. Question asks students to think about a problem. Answer then offers the solution to the problem. Finally, an Explanation section gives the reasoning behind the solution. As with the Key Idea sections, students should make sure they understand these lessons before they proceed. Sections usually begin with a Key Idea screen, and Question/Answer/Explanation is always a sequence. Otherwise there is no set order for the sequencing of these screens in a Guided Reading.

Practice Quiz

Each Practice Quiz consists of ten multiple-choice questions. Answers to the multiple-choice questions appear at the back of the book.

CREDITS

The *Study Guide* accompanies *For All Practical Purposes*, Fifth Edition, by the Consortium for Mathematics and Its Applications (COMAP) and published by W. H. Freeman and Company.

A number of individuals and groups have been involved in developing this *Study Guide.* Guided reading materials were written by Dan Reich of Temple University and based on similar materials developed for the previous edition by Dr. Reich and by Eli Passow, also of Temple University.

Practice Quiz questions and answers were written by John Emert and Kay Meeks Roebuck of Ball State University. Content design and development of the *Study Guide* for the previous edition was overseen by Sandra H. Savage of Orange Coast Community College.

The publisher wishes to thank all of these individuals, as well as the authors of the textbook and the staff of COMAP for their ideas and contributions.

The technical application design and development for *FAPP Interactive* was done by Infon, Inc.

The authors of the *For All Practical Purposes*, Fifth Edition, textbook are as follows:

Project Director: Solomon Garfunkel, Consortium for Mathematics and Its Applications

Contributing Authors: *Part I, Management Science*—Joseph Malkevitch, York College, CUNY; Rochelle Meyer, Nassau Community College; Walter Meyer, Adelphi University.

Part II, Statistics: The Science of Data—David S. Moore, Purdue University.

Part III, Coding Information—Joseph Gallian, University of Minnesota, Duluth.

Part IV, Social Choice and Decision Making—Steven J. Brams, New York University; Bruce P. Conrad, Temple University; Alan D. Taylor, Union College.

Part V, On Size and Shape—Paul J. Campbell, Beloit College.

Major funding for the telecourse was provided by the Annenberg/CPB Project. Additional funding was provided by the Carnegie Corporation of New York. The Alfred P. Sloan Foundation funded the development of the textbook.

CHAPTER 1

STREET NETWORKS

CHAPTER OBJECTIVES

Check off these skills once you feel you have mastered them.

- ❑ Determine by observation if a graph is connected.
- ❑ Identify vertices and edges of a given graph.
- ❑ Construct the graph of a given street network.
- ❑ Determine by observation the valence of each vertex of a graph.
- ❑ Define an Euler circuit.
- ❑ List the two conditions for the existence of an Euler circuit.
- ❑ Determine whether a graph contains an Euler circuit.
- ❑ If a graph contains an Euler circuit, list one such circuit by identifying the order of vertices in the circuit's path.
- ❑ If a graph does not contain an Euler circuit, add a minimum number of edges to eulerize the graph.
- ❑ Identify management science problems whose solutions involve Euler circuits.

GUIDED READING

Introduction

[text p. 3]

key idea

The management of a large and complex system requires careful planning and problem-solving. In this chapter, we focus on an important management issue, one that occurs frequently in a variety of forms: the problem of traversing the network as efficiently and with as little redundancy as possible. Solutions involve mathematical principles as well as practical considerations.

Euler Circuits

[text pp. 3–7]

key idea

The problem of finding an optimal route for checking parking meters or delivering mail can be modeled abstractly as finding a best path through a graph that includes every edge.

question

Represent the street network of mailboxes to be serviced for delivery as a graph. (The triangles represent mailboxes.)

answer

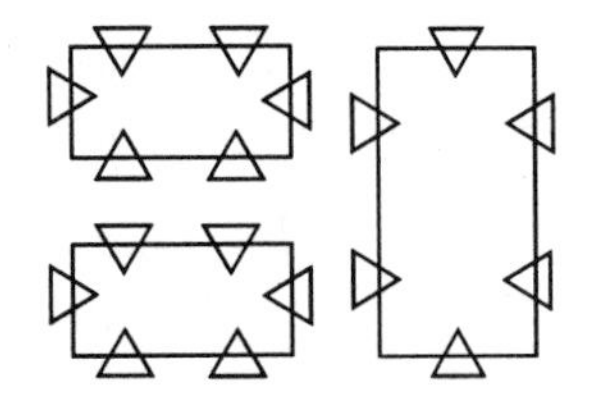

explanation

Start with the basic street network. Without the mailboxes it looks like this:

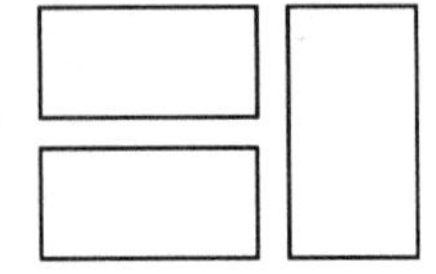

explanation

Now replace each intersection or corner with a vertex. Represent these with circles like this:

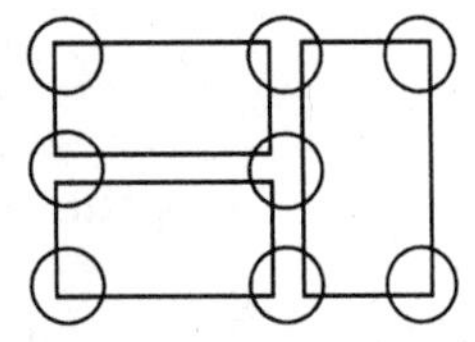

By replacing each row of mailboxes with an edge, the graph in the answer is made.

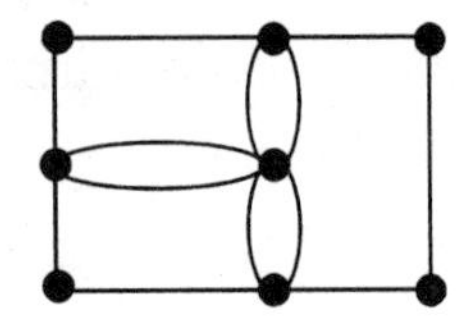

key idea

A path through a graph is a circuit if it starts and ends at the same vertex. A circuit is an Euler circuit if it covers each edge exactly once.

question

Draw an Euler circuit of the graph for the mailbox network.

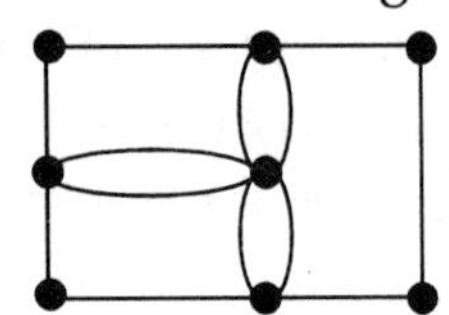

answer

This is one possible Euler circuit for the graph. There are many correct answers.

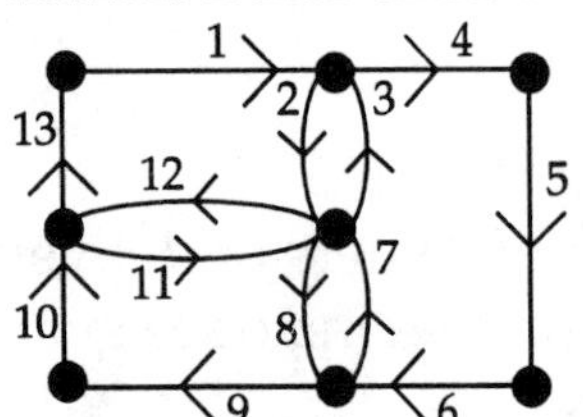

explanation

Starting and ending with the upper left-hand corner, this circuit covers each edge exactly once. Any circuit drawn that meets these conditions (1. starts and ends at the same vertex, and 2. covers each edge exactly once) is an Euler circuit.

Finding Euler Circuits

[text pp. 8–11]

key idea

According to Euler's theorem, a connected graph has an Euler circuit if the valence at each vertex is an even number. If any vertex has an odd valence, there cannot be an Euler circuit.

question

For each of these graphs, find the valence of each vertex. Which graph has an Euler circuit?

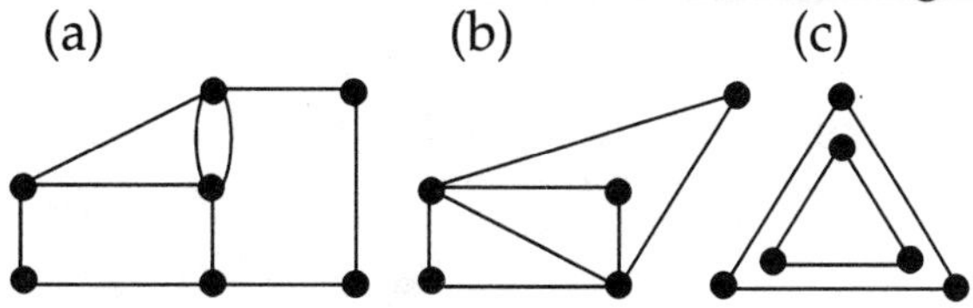

answer

(a) does not have an Euler circuit; (b) does have an Euler circuit; (c) does not have an Euler circuit.

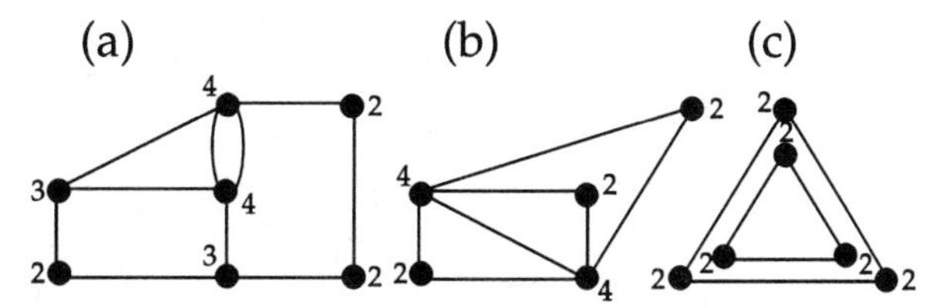

explanation

(a) has two odd vertices, so it cannot have an Euler circuit. (b) has no odd vertices and is connected, so it must have an Euler circuit. (c) has even valences but the graphs are not connected, so it cannot have an Euler circuit. Euler circuits are easy to find.

key idea

In finding an Euler circuit, never "disconnect" the graph by using an edge that is the only link between two parts of the graph not yet covered.

key idea

Here, we are looking for an Euler circuit. Steps 1, 2, and 3 (FBAC) have been completed, and now we must decide where to proceed at vertex C. Proceeding to E "disconnects" EF from CB, CD, and DB. Proceeding to B is permissible. Proceeding to D is also permissible.

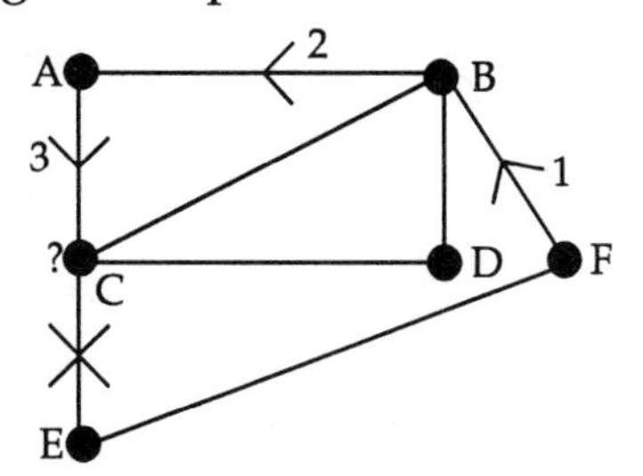

Circuits with Reused Edges

[text pp. 12–18]

key idea

If a graph has odd vertices, then any circuit must reuse at least one edge. The Chinese postman problem involves finding a circuit that reuses as few edges as possible.

key idea

This is a graph that contains odd vertices. One possible circuit follows the sequence of vertices ADCABCADECDA. This circuit reuses four edges: AD (twice), AC, and CD. This is not the circuit that reuses the fewest edges for this graph.

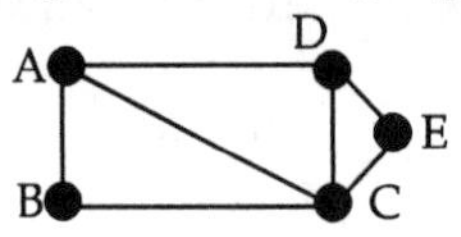

question

Solve the Chinese postman problem for this graph—that is, find an Euler circuit of the graph that reuses the fewest edges.

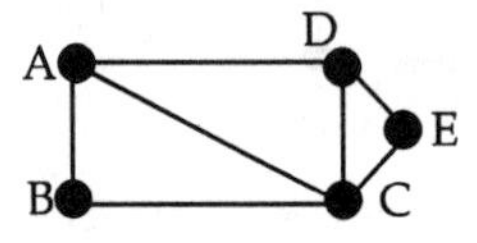

answer

One solution would be to start at A, then follow the sequence ADCBACEDA.

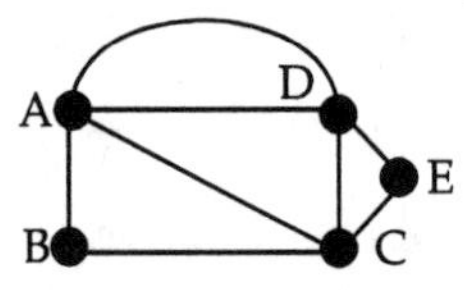

explanation

The circuit ADCBACEDA reuses only one edge, AD. Since the graph does not have an Euler circuit, the best possible result is a circuit that reuses only one edge.

key idea

Reusing an edge that joins two vertices is like adding a new edge between those vertices. Adding new edges for a circuit to produce an Euler circuit of a graph is called eulerizing the original graph.

question

Eulerize this graph.

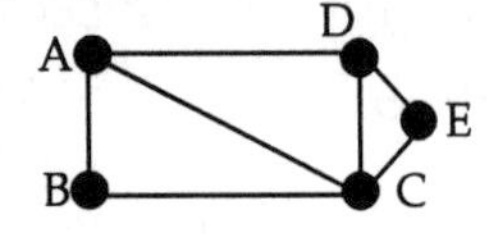

answer

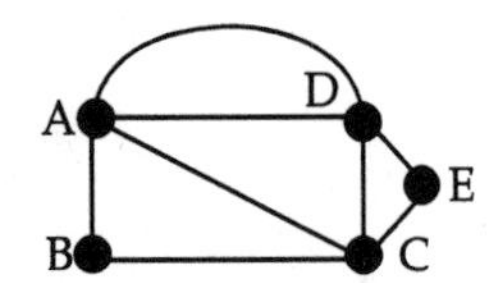

explanation

This is the eulerization of the circuit offered in the answer to the previous question, with the edge AD reused once. This explains the added edge joining vertices A and D.

key idea

An Euler circuit on an eulerized graph can be "squeezed" into a circuit on the original graph by reusing precisely those edges that correspond to the edges added in the eulerization.

key idea

A systematic way to produce a good eulerization of a certain specialized "rectangular" graph is called the "edge walker" technique.

question

Using the "edge walker" technique, eulerize this 2-by-3 rectangular graph and this 2-by-4 rectangular graph.

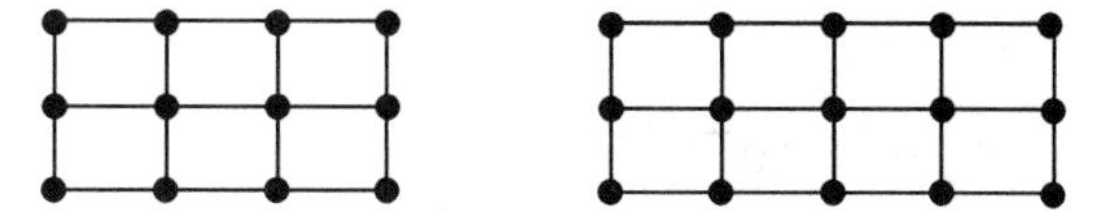

answer

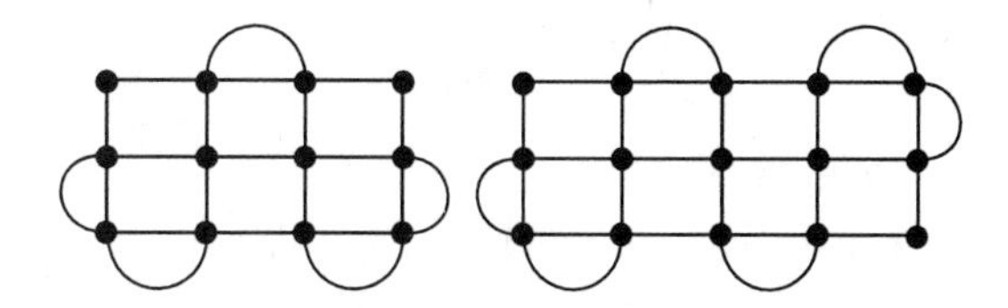

explanation

For each graph, starting with the upper left-hand corner of the graph and traveling clockwise around the boundary of the rectangle, connect each odd vertex you encounter to the next vertex using a new edge.

Circuits with More Complications

[text pp. 18–20]

key idea

Applying the techniques described in this chapter to specific real tasks such as collecting garbage and reading electric meters results in complications that require modification of theories. Types of complications include one-way streets, multi-lane roads, obstructions, and a wide variety of human factors.

PRACTICE QUIZ

1. What is the valence of vertex B in the graph below?

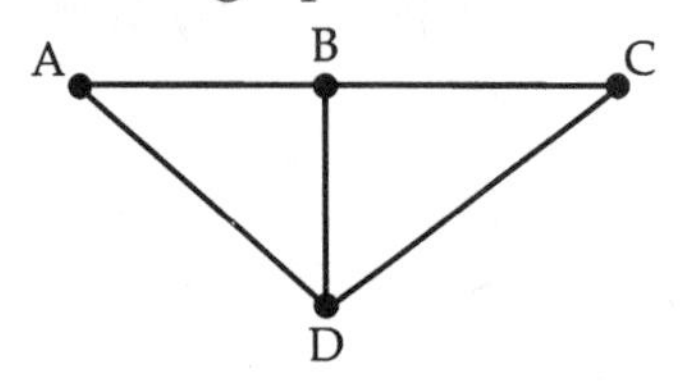

   a. 2
   b. 3
   c. 4

2. A graph is connected only if
   a. every vertex has an even valence.
   b. for every pair of vertices there is a path in the graph connecting these vertices.
   c. it has an Euler circuit.

3. For the graph below, which statement is correct?

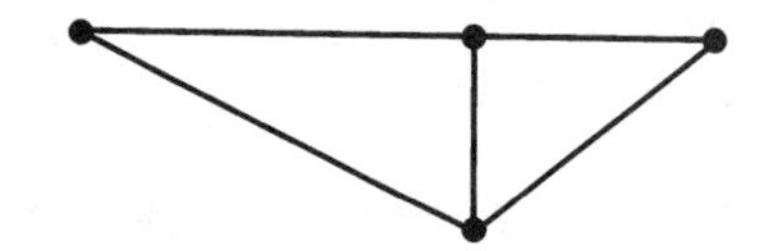

   a. The graph has an Euler circuit.
   b. One new edge is required to eulerize this graph.
   c. Two new edges are required to eulerize this graph.

4. For which of the situations below is it most desirable to find an Euler circuit or an efficient eulerization of the graph?
   a. checking all the fire hydrants in a small town
   b. checking the pumps at the water treatment plant in a small town
   c. checking all the water mains in a small town

5. Consider the path represented by the sequence of numbered edges on the graph below. Which statement is correct?

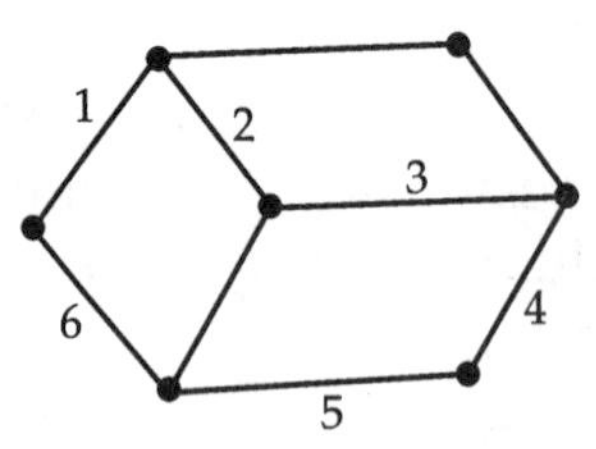

   a. The sequence of numbered edges forms an Euler circuit.
   b. The sequence of numbered edges traverses each edge exactly once, but is not an Euler circuit.
   c. The sequence of numbered edges forms a circuit, but not an Euler circuit.

6. What is the minimum number of duplicated edges needed to create a good eulerization for the graph below?

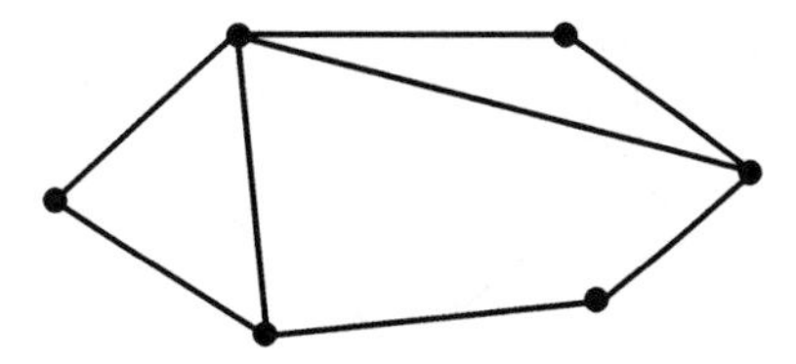

   a. 4 edges
   b. 3 edges
   c. 2 edges

7. Suppose the edges of a graph represent streets along which a postal worker must walk to deliver mail. Why would a route planner wish to find an Euler circuit or an efficient eulerization of this graph?
   a. to minimize the amount of excess walking the carrier needs to do
   b. to determine where postal drop boxes should be placed
   c. both a. and b.

8. If a graph has 10 vertices of odd valence, what is the absolute minimum number of edges that need to be added or duplicated to eulerize the graph?
   a. 5
   b. 10
   c. 0

9. Which option best completes the following analogy: A circuit is to a path as
   a. a vertex is to an edge.
   b. a digraph is to a graph.
   c. operations research is to management science.

10. Which of the following statements is true?
    I. If a graph is connected and has only even valences, then it has an Euler circuit.
    II. If a graph has an Euler circuit, then it must be connected and have only even valences.
    a. Only I is true.
    b. Only II is true.
    c. Both I and II are true.

CHAPTER 2

VISITING VERTICES

CHAPTER OBJECTIVES

Check off these skills once you feel you have mastered them.

- ❑ Write in your own words the definition of a Hamiltonian circuit.
- ❑ Explain the difference between an Euler circuit and a Hamiltonian circuit.
- ❑ Identify a given application as being an Euler circuit problem or a Hamiltonian circuit problem.
- ❑ Calculate $n!$ for a given value of n.
- ❑ Apply the formula $(n-1)!/2$ to calculate the number of Hamiltonian circuits in a graph with a given number of vertices.
- ❑ Define the term algorithm.
- ❑ Explain the term heuristic algorithm and list both an advantage and a disadvantage of using this algorithm.
- ❑ Discuss the difficulties inherent in the application of the brute force method for finding the shortest-route Hamiltonian circuit.
- ❑ Describe the steps in the nearest-neighbor algorithm.
- ❑ Find an approximate solution to the Traveling Salesman Problem by applying the nearest-neighbor algorithm.
- ❑ Describe the steps in the sorted-edges algorithm.
- ❑ Find an approximate solution to the Traveling Salesman Problem by applying the sorted-edges algorithm.
- ❑ Explain the difference between a graph and a tree.
- ❑ Determine from a weighted-edges graph a minimum-cost spanning tree.
- ❑ Identify the critical path on an order-requirement digraph.
- ❑ Find the earliest possible completion time for a collection of tasks by analyzing its critical path.
- ❑ Explain the difference between a graph and a directed graph.

GUIDED READING

Introduction

[text p. 31]

key idea

Using a salesman's tour from city to city as a model, we investigate efficient ways to traverse a graph while visiting each vertex only once, along with some related problems. Our goal usually is to find a tour which is as quick, or as cheap as possible.

Hamiltonian Circuits

[text pp. 31–38]

key idea

A Hamiltonian circuit of a graph visits each vertex exactly once, and returns to the starting point. There is no simple way to determine if a graph has a Hamiltonian circuit, and it can be hard to construct one.

question

Find a Hamiltonian circuit for this graph starting at A. (Remember: unlike the Euler circuit, it is not necessary to traverse every edge.)

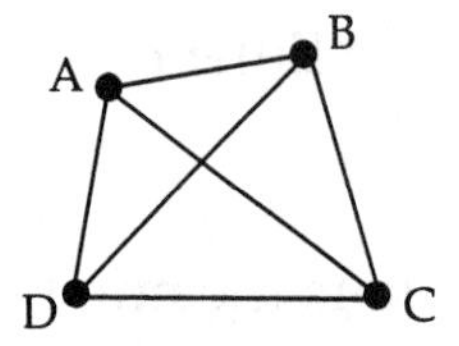

answer

ABCDA, ACDBA, and ADBCA—as well as their reversals, ADCBA, ABDCA, and ACBDA—are all Hamiltonian circuits.

explanation

These are the six possible Hamiltonian circuits starting from A. If you got a different answer, did you add a vertex where the diagonals cross? You shouldn't have. Since a vertex has not been indicated, edges AC and BD do not actually intersect. (You may think of AC as representing a highway overpass that is on a different level from edge BD.)

question

Does this graph have a Hamiltonian circuit?

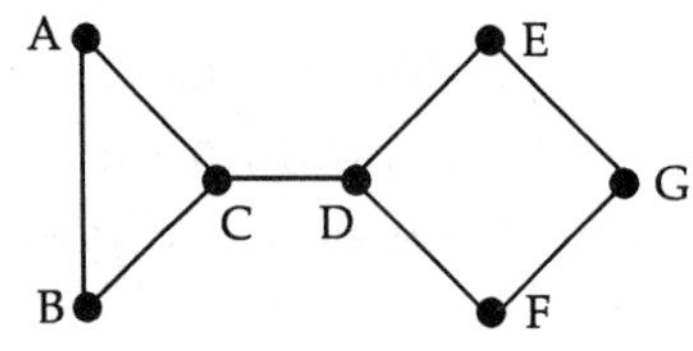

answer

No, the graph does not have a Hamiltonian circuit.

explanation

The edge CD divides the graph into two parts. If you start the tour at a vertex in one part and then cross CD, you cannot get back to the starting vertex without crossing CD again.

key idea

A minimum-cost Hamiltonian circuit is one with the lowest possible sum of the weights of its edges.

key idea

One very important class of graphs, the complete graphs, automatically have Hamiltonian circuits. Graphs may fail to have Hamiltonian circuits for a variety of reasons.

question

Which of these graphs has a Hamiltonian circuit?

(a)

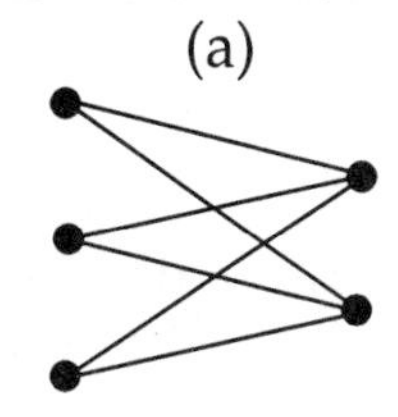

(b)

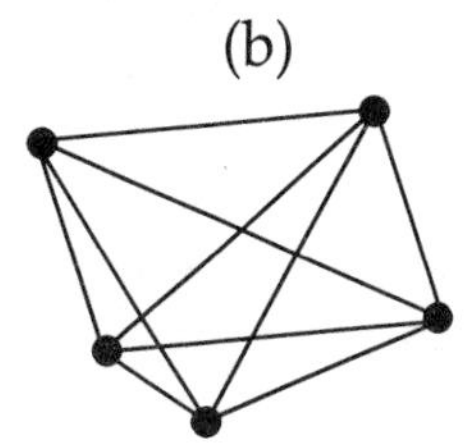

answer

(a) does not have a Hamiltonian circuit, (b) does have a Hamiltonian circuit.

explanation

(a) is from a family of graphs known not to have Hamiltonian circuits: it was constructed with vertices on two parallel vertical columns with one column having more vertices than the other. (b) is from a family of graphs known to have Hamiltonian circuits—the family of complete graphs; every pair of vertices is joined by an edge.

question

Determine how many Hamiltonian circuits there are for a complete graph with 10 vertices.

answer

181,440 possible Hamiltonian circuits.

explanation

If a complete graph has n vertices, then there are $(n-1)!/2$ Hamiltonian circuits.

$$\text{For } n = 10, \text{ we get } \frac{(10-1)!}{2} = \frac{9\times 8\times 7\times 6\times 5\times 4\times 3\times 2\times 1}{2} = \frac{362{,}880}{2} = 181{,}440$$

key idea

The brute force method is one algorithm that can be used to find a minimum-cost Hamiltonian circuit, but it is not a very practical method for a large problem.

The Traveling Salesman Problem

[text pp. 38–39]

key idea

Remember the earlier statement:

"A minimum-cost Hamiltonian circuit is one with the lowest possible sum of the weights of its edges."

The problem of finding this minimum-cost Hamiltonian circuit is called the traveling salesman problem (TSP). It is a common goal in the practice of management science.

Strategies for Solving the TSP

[text pp. 39–43]

key idea

We use a variety of heuristic (or "fast") algorithms to find solutions to the TSP. Some are very good, even though they may not be optimal. Heuristic algorithms come close enough to giving optimal solutions to be important for practical use.

key idea

The nearest-neighbor algorithm repeatedly selects the closest neighboring vertex not yet visited in the circuit, and returns to the initial vertex at the end of the tour.

question

Using the nearest-neighbor algorithm, finish finding a Hamiltonian circuit for this graph. We started at vertex C, proceeded to D, and then to B. When you've finished, calculate the total.

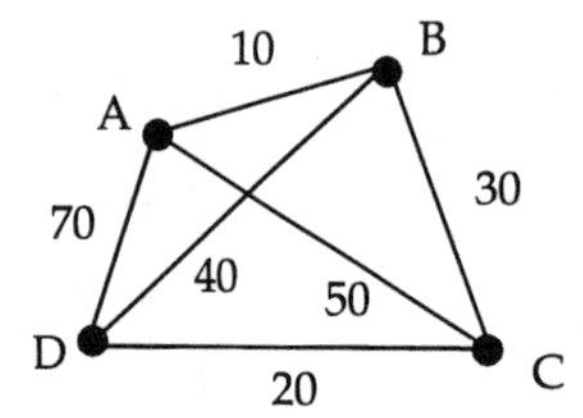

answer

CDBAC is the complete grid. The total cost is 120.

explanation

From B, the closest neighbor that has not yet been visited is A. All the vertices have thus been visited and you can return to starting point C. The total of the weights of the edges in the path is 120.

key idea

The sorted-edges algorithm (which, like nearest neighbor, is a greedy algorithm) is another heuristic algorithm that can lead to a solution that is close to optimal.

question

For this graph, what are the first three edges chosen according to the sorted-edges algorithm? Complete the circuit and calculate the total cost.

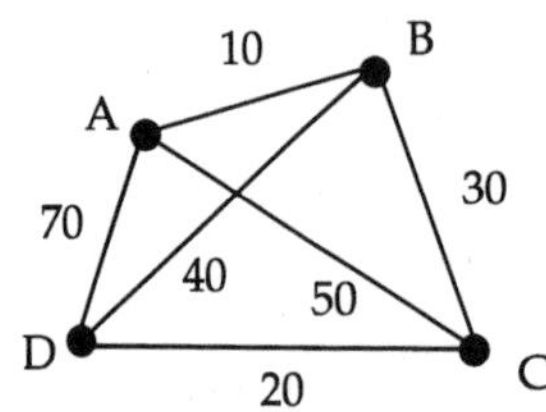

answer

The first three edges, in order, are AB, CD, BC. The complete circuit is ABCDA, which has a total cost of 130.

explanation

The edges AB, CD, and BC are the cheapest, and they do not close a loop or meet at a single vertex. This forces the choice of the expensive edge, DA, to return to the starting vertex. The cost of all the edges in the tour adds up to 130.

Minimum-Cost Spanning Trees

[text pp. 43–49]

key idea

We can use a spanning tree to connect all the vertices of a graph to each other with no redundancy (e.g., for a communications network.)

question

With (a) as our original graph, which of the graphs shown in bold represent trees and/or spanning trees?

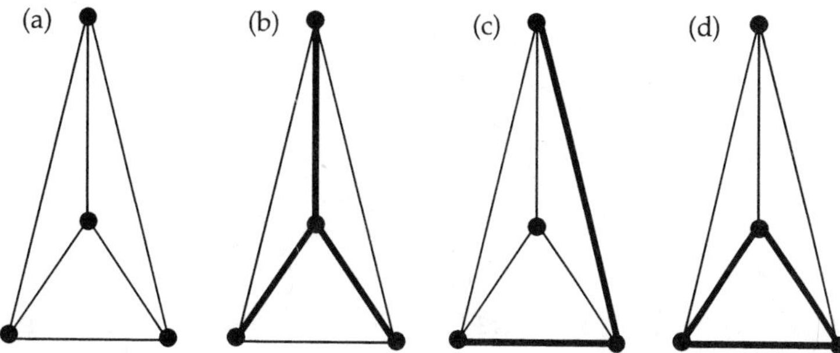

answer

(b) is a tree and a spanning tree; (c) is a tree, but not a spanning tree; and (d) is neither a tree nor a spanning tree.

explanation

(b) is a tree because it is a connected graph with no circuits, and it is a spanning tree because it includes all the vertices of the original graph; (c) is a tree because it is a connected graph with no circuits, but is not a spanning tree because it does not include all the vertices of the original graph; and (d) contains a circuit, so it cannot be a tree, and therefore cannot be a spanning tree.

key idea

Two spanning trees for graph (a) are shown in (b) and (c).

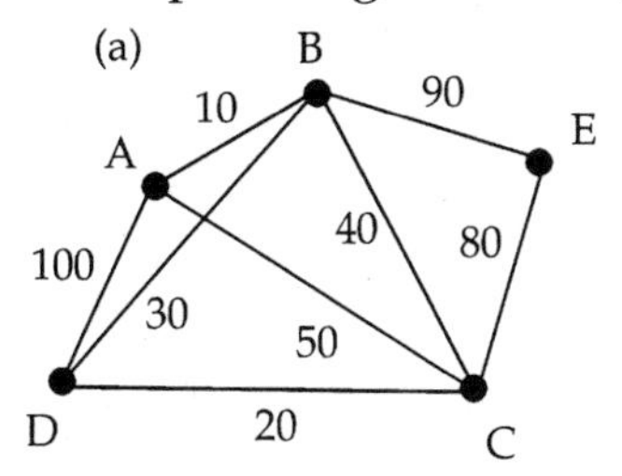

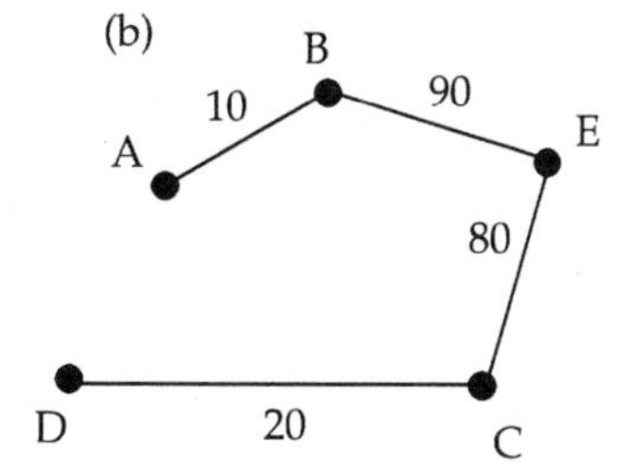

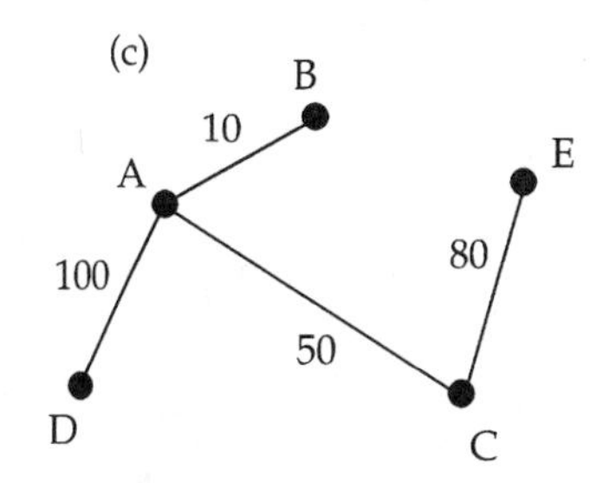

Each is an example of a minimal subgraph connecting all the vertices in the original. In each case, removal of any edge will disconnect the graph.

key idea

A minimum-cost spanning tree is most economical. Kruskal's algorithm produces one quickly.

question

Using this graph again, apply Kruskal's algorithm to find a minimum-cost spanning tree.

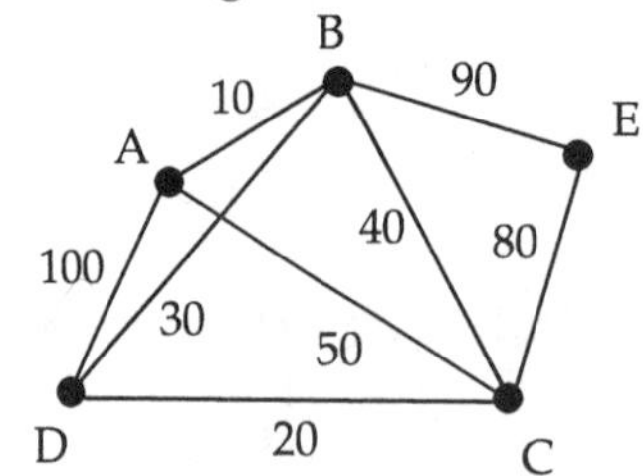

answer

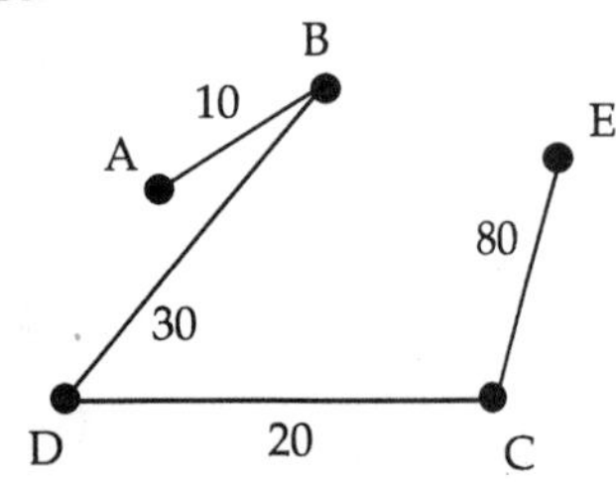

explanation

AB, DC, and BD are the cheapest edges and are chosen first. The next cheapest ones are BC and AC, but these would close loops. EC is the next in line, and that completes the tree.

Critical-Path Analysis

[text pp. 50–54]

key idea

The necessary arrangement of tasks in a complex job can be represented in a digraph (directed graph), with arrows showing the order requirements.

question

For this order-requirement digraph, are the statements true or false?

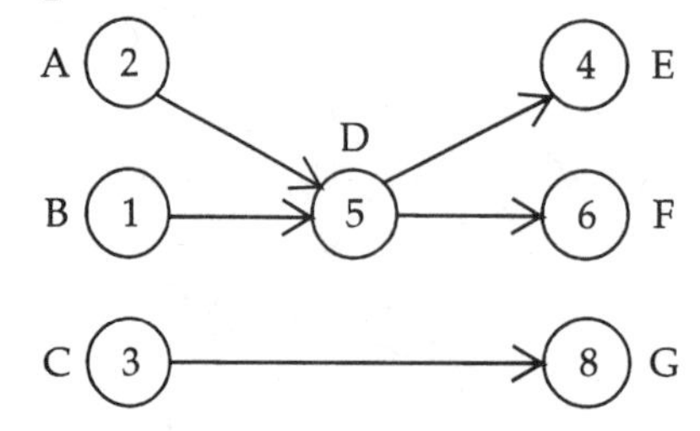

a) A must be done before B.
b) B must be done before D.
c) A must be done before G.
d) B need not be done before E.
e) C need not be done before E.

answer

a) A must be done before B. **False** A is independent of B.
b) B must be done before D. **True** B precedes D.
c) A must be done before G. **False** No connection between A and G.
d) B need not be done before E. **False** B precedes D, so B must precede E.
e) C need not be done before E. **True** No connection between C and E.

explanation

A is independent of B, since no sequence of arrows connects them. Nor is there any connection between A and G, or C and E. However, B must precede D, and therefore B must also precede E.

question

Find the critical path for this digraph.

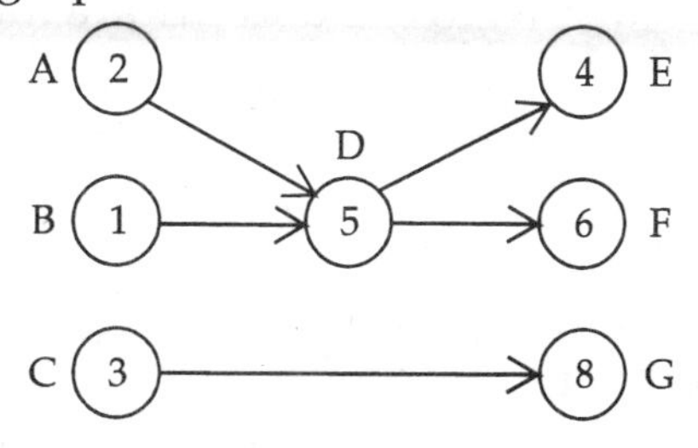

answer

ADF.

explanation

ADF is the path through the digraph with the greatest total time: $2 + 5 + 6 = 13$.

PRACTICE QUIZ

1. Which of the following describes a Hamiltonian circuit for the graph below?

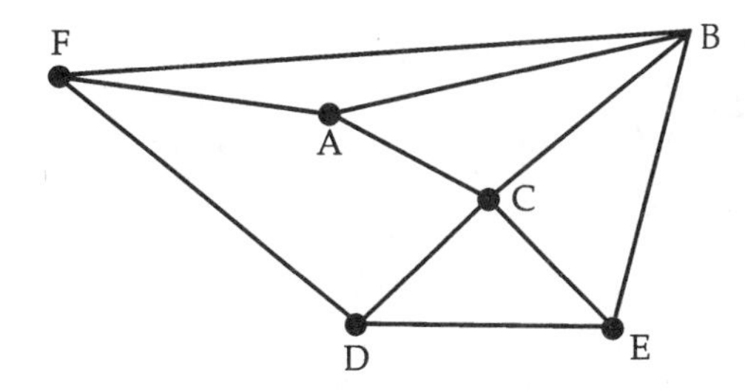

a. ABCEDCBFDA
b. ABCDEFA
c. ACBEFDA

2. Using the nearest-neighbor algorithm and starting at vertex A, find the cost of the Hamiltonian circuit for the graph below.

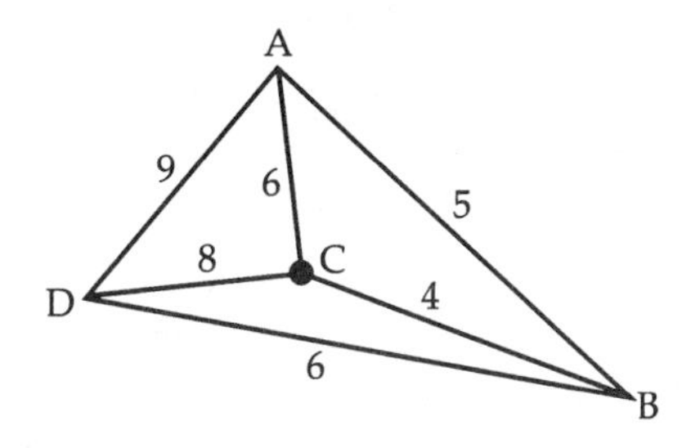

a. 25
b. 26
c. Another answer

3. Using the sorted-edges algorithm, find the cost of the Hamiltonian circuit for the graph below.

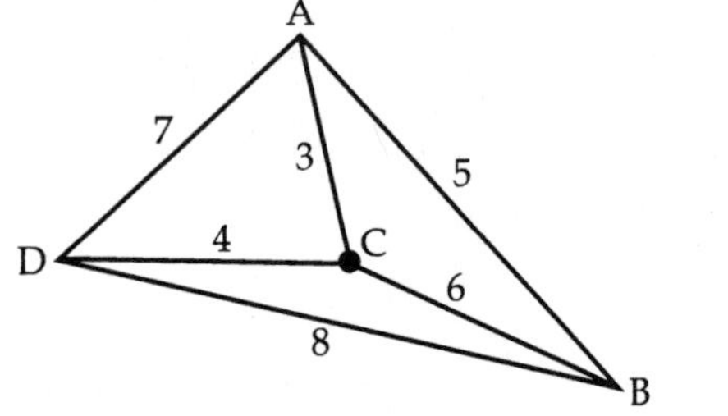

a. 20
b. 22
c. 18

4. Using Kruskal's algorithm, find the minimum-cost spanning tree for the graph below. Which statement is true?

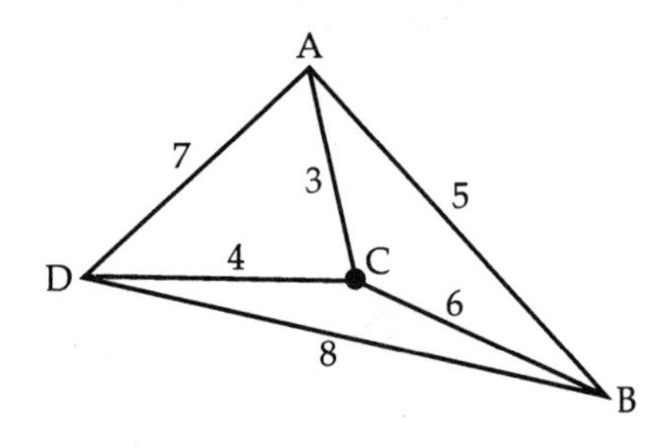

a. Edges AC and BD are included.
b. Edges AC and AB are included.
c. Edges CD and BD are included.

5. Which of the following statements are true?
   I: It can be proved that Kruskal's algorithm always produces an optimal solution.
   II: If a graph has five vertices, its minimum-cost spanning tree will have four edges.
   a. Only I is true.
   b. Only II is true.
   c. Both I and II are true.

6. What is the critical path for the following order-requirement digraph?

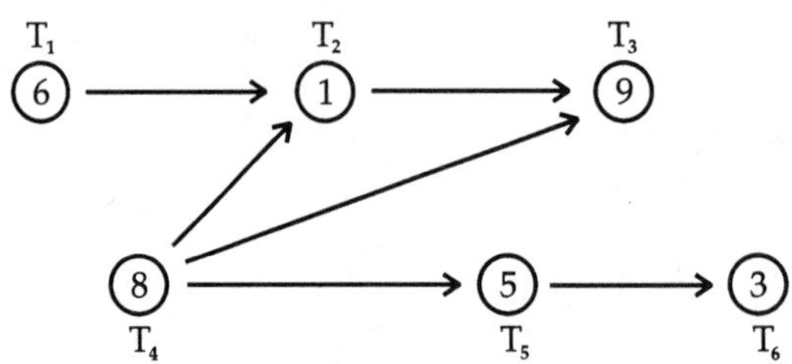

a. T4, T2, T3
b. T4, T3
c. T1, T2, T3 and T4, T5, T6 are both critical paths.

7. What is the earliest completion time (in minutes) for a job with the following order-requirement digraph?

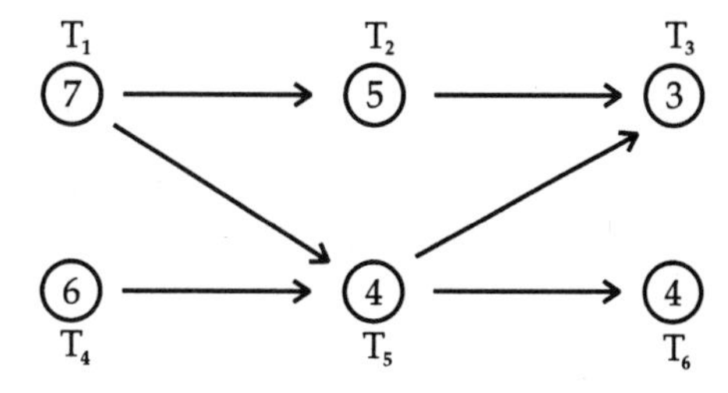

   a. 13 minutes
   b. 14 minutes
   c. 15 minutes

8. Which of the following statements are true?
   I: Trees never contain circuits.
   II: Trees are always connected and always include all the vertices of the larger graph.
   a. Only I is true.
   b. Only II is true.
   c. Neither I nor II is true.

9. Five small towns decide to set up an emergency communication system by connecting to each other with fiber optic cable. Which technique is most likely to be useful in helping them do this as cheaply as possible?
   a. finding an Euler circuit on a graph
   b. finding a Hamiltonian circuit on a graph
   c. finding a minimum-cost spanning tree on a graph

10. Terry has three types of bread, four kinds of deli meat, and three types of cheese. How many different sandwiches could Terry make?
    a. fewer than 15
    b. between 15 and 40
    c. more than 40

CHAPTER 3

PLANNING AND SCHEDULING

CHAPTER OBJECTIVES

Check off these skills once you feel you have mastered them.

- ❑ State the assumptions for the scheduling model.
- ❑ Compute the lower bound on the completion time for a list of independent tasks on a given number of processors.
- ❑ Describe the list-processing algorithm.
- ❑ Apply the list-processing algorithm to schedule independent tasks on identical processors.
- ❑ For a given list of independent tasks, compare the total task time using the list-processing algorithm for both the non-sorted list and also a decreasing-time list.
- ❑ When given an order-requirement digraph, apply the list processing algorithm to schedule a list of tasks subject to the digraph.
- ❑ Explain how a bin-packing problem differs from a scheduling problem.
- ❑ Given an application, determine whether its solution is found by the list-processing algorithm or by one of the bin-packing algorithms.
- ❑ Discuss advantages and disadvantages of the next-fit bin-packing algorithm.
- ❑ Solve a bin-packing problem by the non-sorted next-fit algorithm.
- ❑ Solve a bin-packing problem by the decreasing-time next-fit algorithm.
- ❑ Discuss advantages and disadvantages of the first-fit bin-packing algorithm.
- ❑ Apply the non-sorted first-fit algorithm to a bin-packing problem.
- ❑ Apply the decreasing-time first-fit algorithm to a bin-packing problem.
- ❑ Discuss advantages and disadvantages of the worst-fit bin-packing algorithm.
- ❑ Find the solution to a bin-packing problem by the non-sorted best-fit algorithm.
- ❑ Find the solution to a bin-packing problem by the decreasing-time worst-fit algorithm.
- ❑ List two examples of bin-packing problems.
- ❑ Create a vertex coloring of a graph, and explain its meaning in terms of assigned resources.
- ❑ Find the chromatic number of a graph.
- ❑ Interpret a problem of allocation of resources with conflict as a graph, and find an efficient coloring of the graph.

GUIDED READING

Introduction

[text p. 72]

We consider efficient ways to schedule a number of related tasks, with constraints on the number of workers, machines, space, or time available.

Scheduling Tasks

[text pp. 72–81]

key idea

The **machine-scheduling problem** is to decide how a collection of tasks can be handled by a certain number of **processors** as quickly as possible. We have to respect both order requirements among the tasks and a **priority list.**

key idea

The **list-processing algorithm** chooses a task for an available processor by running through the priority list in order, until it finds the **ready task.**

question

In this digraph, which tasks are ready: (a) as you begin scheduling? (b) if just T_2 and T_3 have been completed?

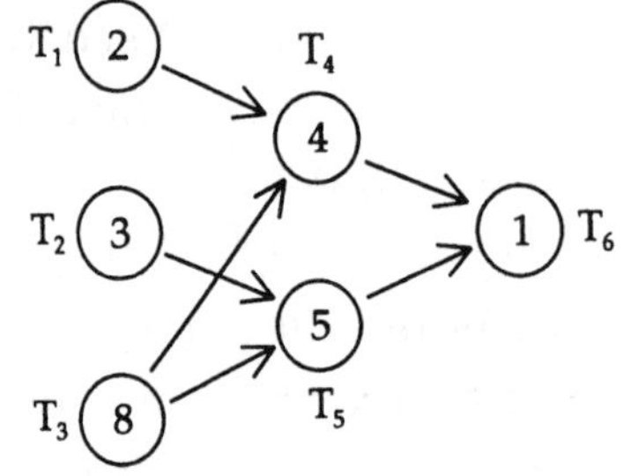

answer

The answer for part (a) is T_1, T_2, T_3 and the answer for part (b) is T_5.

explanation

(a) At the start, only T_1, T_2, and T_3 have no required predecessors.
(b) With T_2 and T_3 completed, T_5 is ready but T_4 must wait for T_1. Also, T_6 must wait for T_4 and T_5.

key idea

Different priority lists can lead to different schedules and completion times.

question

Schedule the tasks in the digraph on two processors with priority list $\{T_1, T_2, T_3, T_4, T_5, T_6\}$ and determine the completion time.

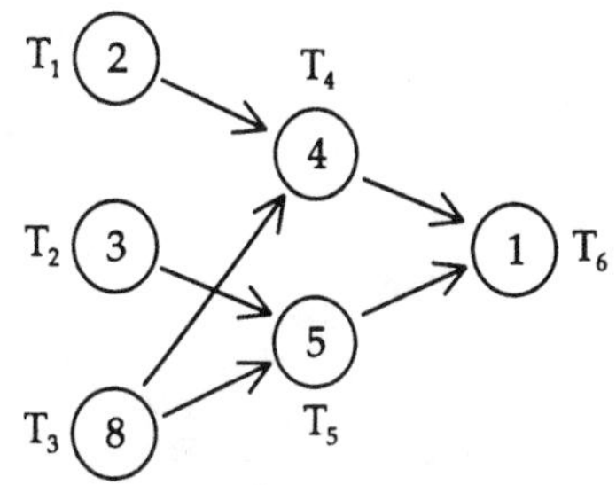

answer

The completion time is 16. This is the schedule:

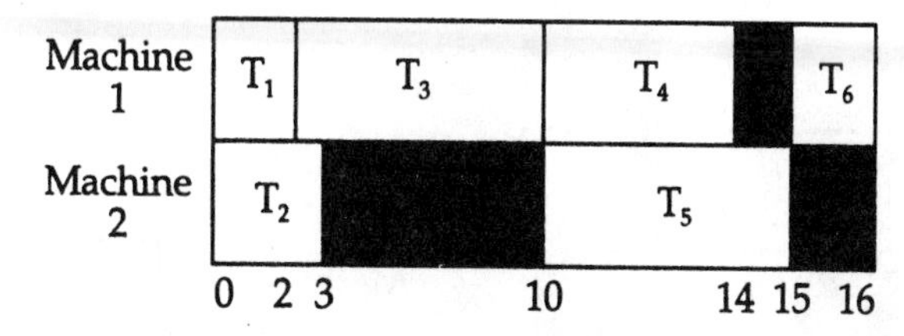

explanation

T_1 and T_2 are first, then T_3. T_4 and T_5 must wait for completion of T_3. Also, T_6 must wait for T_4 and T_5.

question

Schedule the tasks in the digraph on two processors with priority list $\{T_6, T_5, T_4, T_3, T_2, T_1\}$ and determine the completion time.

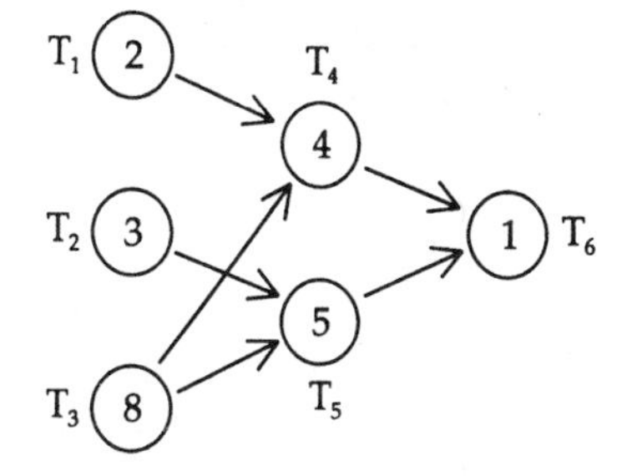

answer

The completion time is 14. This is the schedule:

Machine 1	T_3		T_5		T_6
Machine 2	T_2	T_1		T_4	
	0	3 5	8	12	13 14

explanation

T_3 is the highest priority ready task and gets scheduled first, with T_2 next. After T_2 is done, no task is ready except for T_1, so T_1 is scheduled. When T_3 is done, T_6 is not ready so T_5 is scheduled, followed by T_4 and T_6.

key idea

A schedule is optimal if it has the earliest possible total completion time. For example, a critical path in the order-requirement digraph may determine the earliest completion time.

question

Find a critical path in the digraph.

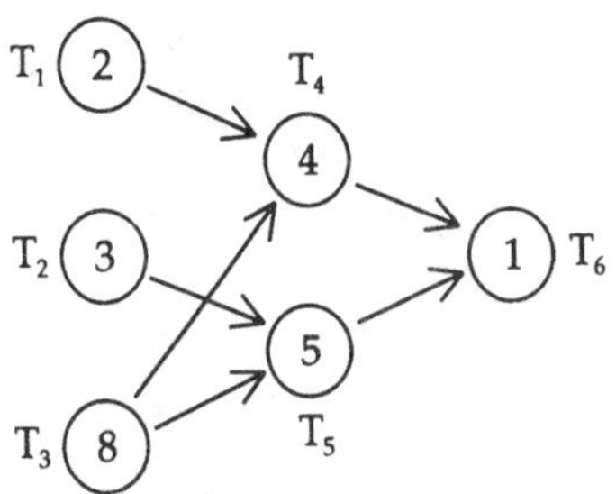

Is one of the schedule constructs optimal? If so, which one?

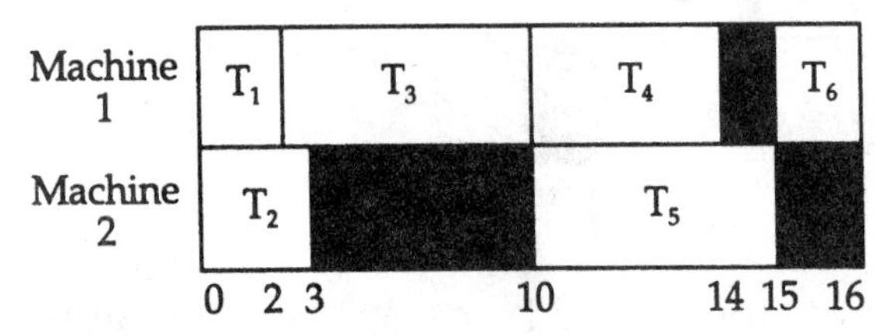

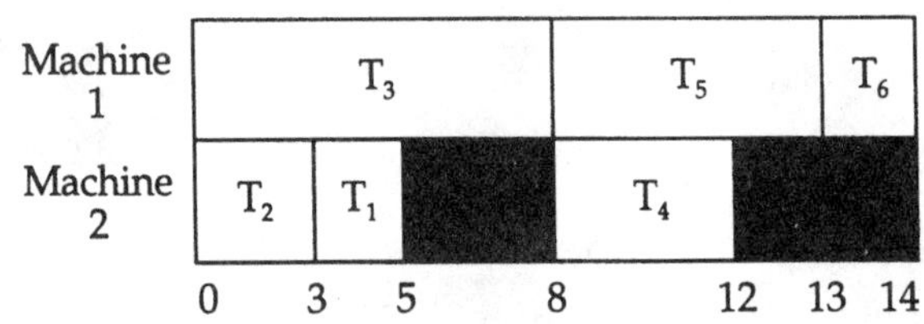

answer

$\{T_3, T_5, T_6\}$ is a critical path of length 14. The second schedule is optimal.

explanation

$\{T_3, T_5, T_6\}$ is the longest path in the digraph, and no scheduling can be completed in less time than the length of this path. Because the second schedule is completed at the same time of 14, it matches the critical path length of 14.

Critical-Path Scheduling

[text pp. 82–85]

key idea

If we can choose or change a priority list, then we have a chance to find an optimal schedule.

key idea

The **critical-path scheduling** algorithm schedules first the tasks in a critical path.

question

Use critical-path scheduling to construct a priority list for the tasks in this digraph.

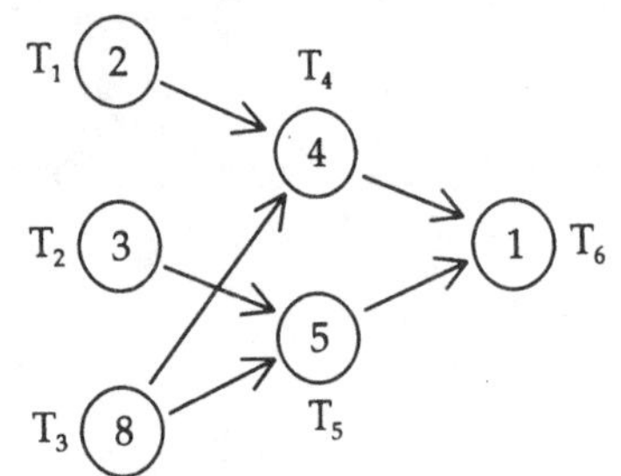

answer

$\{T_3, T_2, T_1, T_5, T_4, T_6\}$

explanation

T_3 is at the head of a critical path. When you remove T_3 and its arrows, T_2 is the head of the remaining critical path $\{T_2, T_5, T_6\}$. Removing T_2 and its arrows makes T_1 the head of $\{T_1, T_4, T_6\}$. Finally, T_5 is the head of $\{T_5, T_6\}$.

question

Construct the schedule for the tasks based on the critical path priority list from the above, $\{T_3, T_2, T_1, T_5, T_4, T_6\}$.

answer

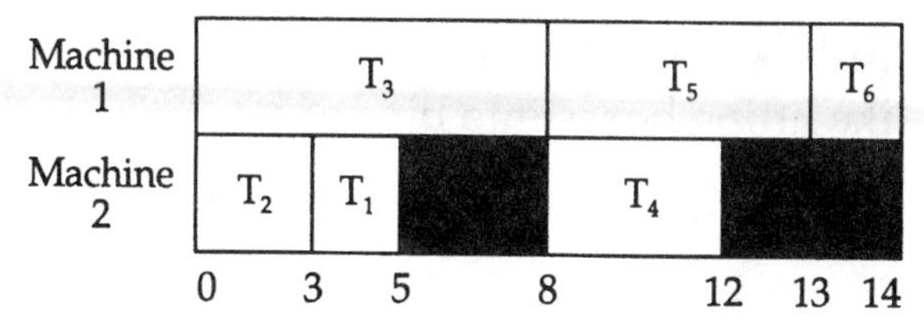

Independent Tasks

[text pp. 86–89]

key idea

When a set of tasks are **independent** (can be done in any order), we have a variety of available algorithms to choose a priority list leading to close-to-optimal scheduling. Some algorithms perform well in the **average-case,** but poorly in the **worst-case.**

question

What is Ronald Graham's worst-case estimate for applying list-processing to any priority list of independent tasks, if the optimal time is 60 hours and you have three processors?

answer

The answer is 100 hours.

explanation

The formula says that the completion time is at most $(2(1/m)T)$, where m is the number of processors and T is the theoretical optimal time.

For $m = 2$, $T = 60$.

This yields $(2 - 1/3)(60) = (1\ 2/3)(60) = 300/3 = 100$

question

The **decreasing-time-list algorithm** schedules the longest tasks earliest. By erasing the arrows in the digraph used throughout this chapter, we obtain a set of independent tasks. We then construct a decreasing-time priority list. Use this list to schedule the tasks. Also, determine the completion time.

answer

The list is {T_3, T_5, T_4, T_2, T_1, T_6}. The schedule leads to a completion time of 12.

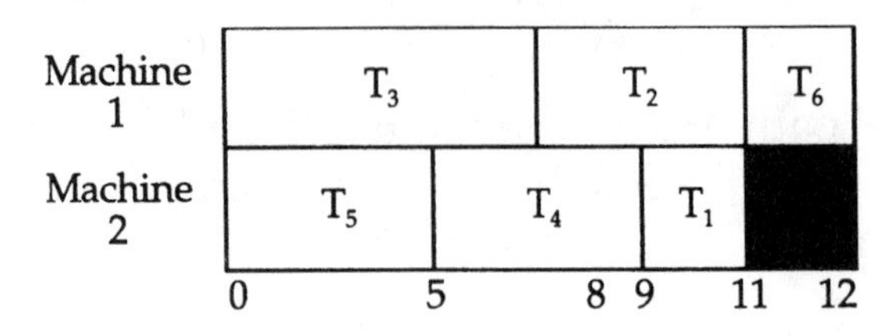

Bin Packing

[text pp. 90–93]

key idea

With the bin-packing problem, we consider scheduling tasks within a fixed time limit, using as few processors as possible. This is like fitting boxes into bins of a certain size—but it is used in a variety of real-world applications.

key idea

We have a variety of heuristic algorithms available to do the packing well if not optimally. Three important algorithms are **next fit** (NF), **first fit** (FF), and **worst fit** (WF).

question

Use the next-fit algorithm to pack boxes of sizes {4, 5, 1, 3, 4, 2, 3, 6, 3} into bins of capacity 8. How many bins are required?

answer

The answer is six bins.

explanation

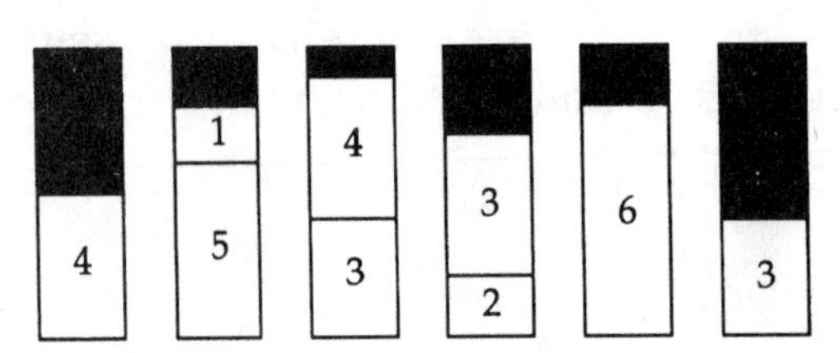

Bin 1 did not have enough space left for the second box, so bin 2 was used. There was enough room in bin 1 for the third box, but the NF heuristic doesn't permit us to go back to earlier bins. Once a bin is opened, it is used as long as the boxes fit—if they don't fit, a new bin is opened.

question

Now use the first-fit algorithm to pack boxes of sizes {4, 5, 1, 3, 4, 2, 3, 6, 3} into bins of capacity 8. How many bins are required?

answer

The answer is five bins.

☛ **explanation**

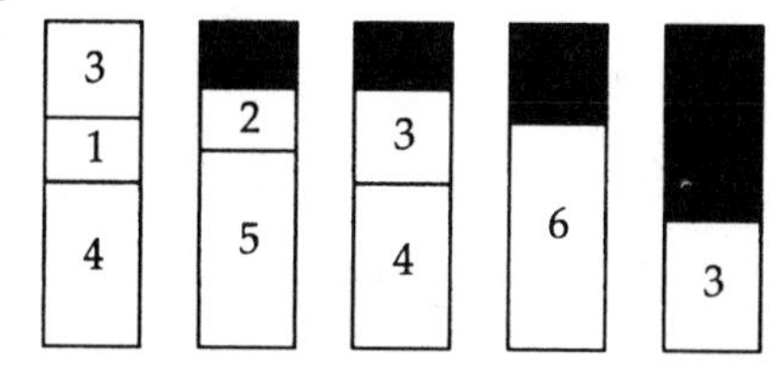

After opening bin 2 for the second box we were able to go back to bin 1 for the next two boxes. The FF heuristic allows us to return to earlier bins while the NF does not.

key idea

WF is like FF in that it permits returning to earlier bins. However, in FF you always start back a the first bin and sequentially search for a bin that will accommodate this weight, while in WF you calculate the unused space in each available bin and select the bin with the maximum room.

key idea

Each of these algorithms can be combined with **decreasing-time** heuristics, leading to the three algorithms next-fit decreasing (NFD), first-fit decreasing (FFD), and worst-fit decreasing (WFD).

question

Now use the first-fit decreasing algorithm to pack the boxes of sizes {4, 5, 1, 3, 4, 2, 3, 6, 3} into bins of capacity 8. How many bins are required?

answer

The answer is four bins.

☛ **explanation**

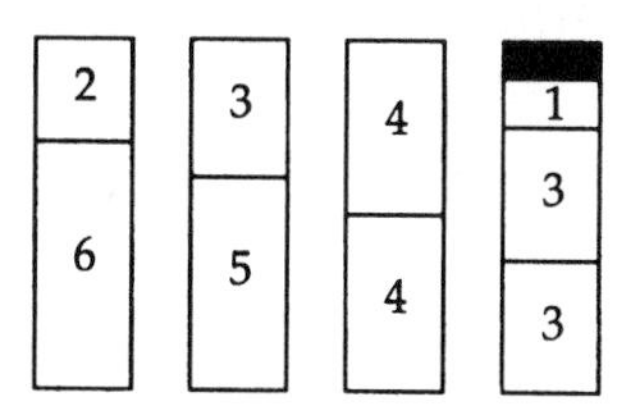

First, rearrange the boxes in size order: {6, 5, 4, 4, 3, 3, 3, 2, 1}.

question

Is any of the algorithms you have used to pack the boxes sized {4, 5, 1, 3, 4, 2, 3, 6, 3} an optimal packing (that is, one using the fewest bins)?

answer

In this case, FFD found the optimal packing. However, neither FFD nor any of the other heuristics discussed in this section will necessarily find the optimal number of bins in an arbitrary problem.

☛ **explanation**

The amount of unused space is obviously less than the capacity of one bin. Therefore, no fewer than four bins could be used to hold all the boxes in this problem.

Resolving Conflict

[text pp. 93–98]

key idea

If we represent items to be scheduled (classes, interviews, etc) as vertices in a graph, then a **vertex coloring** of the graph can be used to assign resources, such as times or rooms, to the items in a conflict-free manner.

key idea

The **chromatic number** of the graph determines the minimum amount of the resource that must be made available for a conflict-free schedule.

Here is a graph with five vertices which is colored using four colors {A, B, C, D}.

question

What is the maximum vertex valence for this graph?

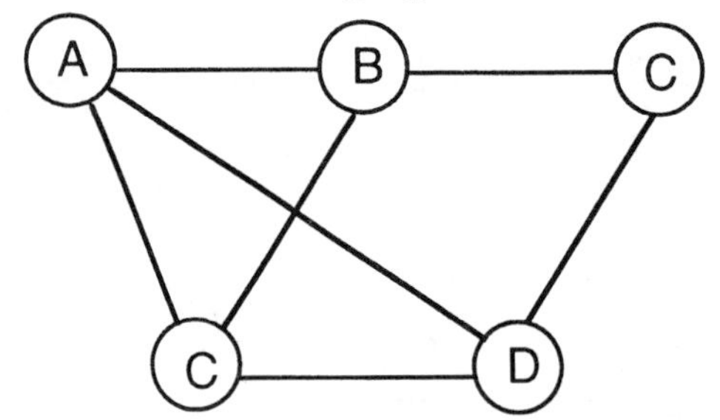

answer

Maximum valence = 3

explanation

The vertex labelled A, for example, is connected to three other vertices, labelled C, B, D. No vertex is connected to four others.

question

Can you find a vertex coloring of the same graph with only three colors?

answer

Brooks' Theorem guarantees that this graph of maximum valence 3 can be vertex-colored with 3 colors. Here is one way to do it.

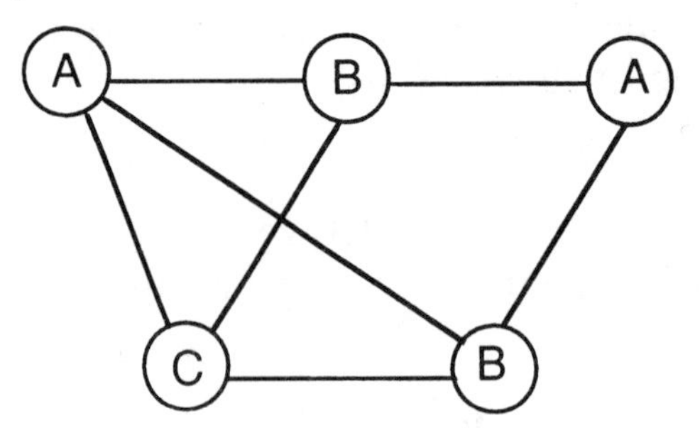

key terms and phrases

- vertex coloring
- chromatic number
- Brooks' theorem

PRACTICE QUIZ

1. Given the order-requirement digraph below (with time given in minutes) and the priority list $T_1, T_2, T_3, T_4, T_5, T_6$, apply the list-processing algorithm to construct a schedule using two processors. How much time is required?

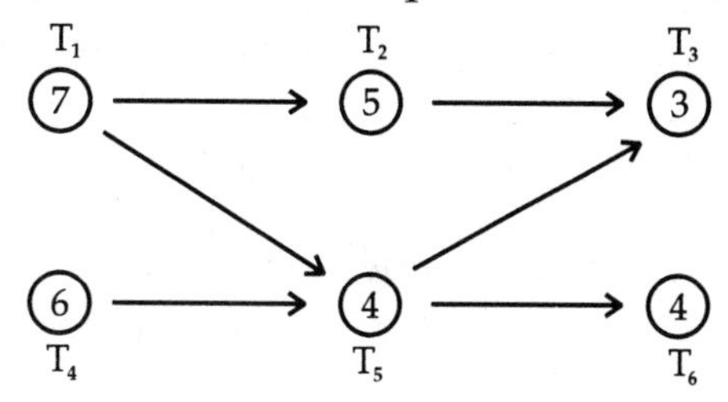

   a. 13 minutes
   b. 15 minutes
   c. 16 minutes

2. Given the order-requirement digraph below (with time given in minutes) and the priority list $T_1, T_2, T_3, T_4, T_5, T_6$, apply the critical-path scheduling algorithm to construct a schedule using two processors. Which task is scheduled first?

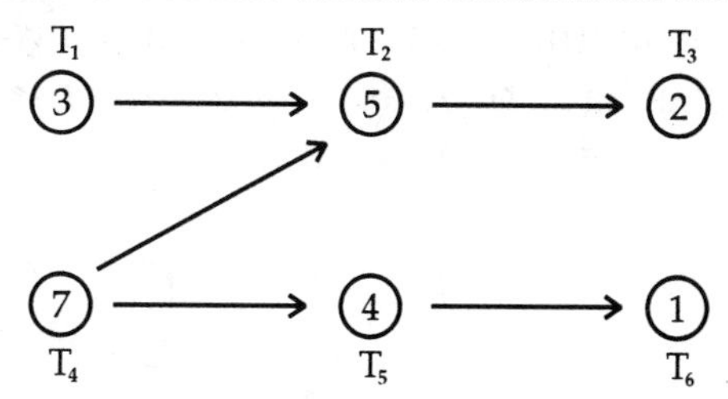

   a. T1
   b. T2
   c. T4

3. The director of a skating show has 25 skaters with varying length numbers to split into three segments, separated by intermissions. This job can be solved using:
   a. the list-processing algorithm for independent tasks
   b. the critical-path scheduling algorithm
   c. the first-fit algorithm for bin packing

4. Use the decreasing-time-list algorithm to schedule the tasks below (times given in minutes) on 2 machines. How much time does the resulting schedule require?
   Tasks: 5, 4, 7, 3, 8, 6, 2, 5, 8
   a. 24 minutes
   b. 25 minutes
   c. more than 25 minutes

5. Use the first-fit algorithm to pack the following weights into bins that can hold no more than 10 pounds. How many bins are required?
   Weights: 5, 4, 7, 3, 8, 6, 2, 5, 8
   a. 5
   b. 6
   c. more than 6

6. Compare the results of the first-fit and the first-fit-decreasing algorithm to pack the following weights into bins that can hold no more than 10 pounds. Which statement is true?

   Weights: 5 , 4, 7, 3, 8, 6, 2, 5, 8

   a. The two algorithms pack the items together in the same way.
   b. The two algorithms use the same number of bins, but group the items together in different ways.
   c. One algorithm uses fewer bins than the other.

7. Find the chromatic number of the graph below:

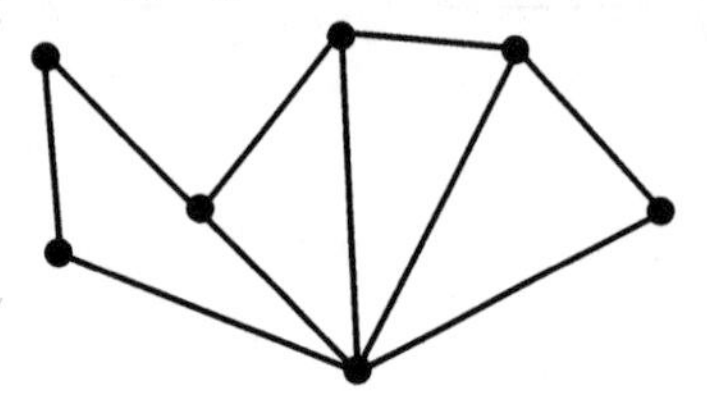

   a. 5
   b. 4
   c. 3

8. An architecture firm must schedule meeting times for its working groups. The following chart indicates which projects have overlapping members for their working groups. Which graph would be used to decide how many different meeting times would be required?

	A	B	C	D	E
A		X		X	
B	X		X		
C		X		X	X
D	X		X		
E			X		

a.

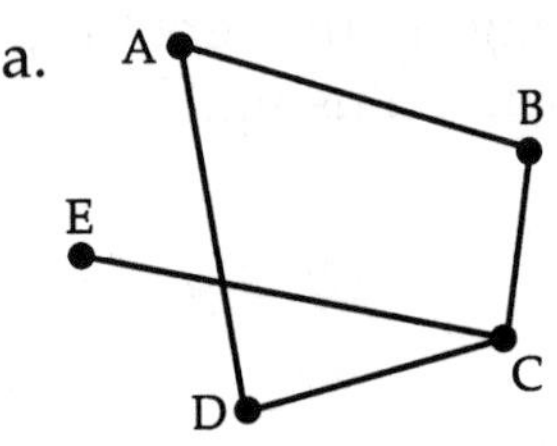

b.

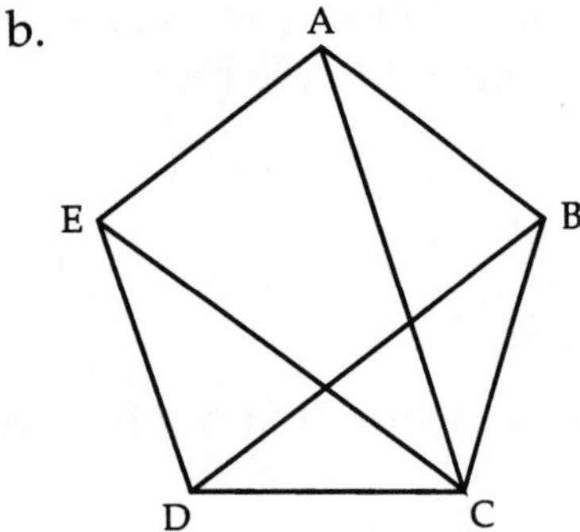

c. neither

9. Which of the following statements is true?

   I: If the time of a task on the critical path of a digraph is shortened by 3 minutes, overall completion time for the job is shortened by 3 minutes.

   II: If additional processors are devoted to a job, the overall time required would be less.

   a. I only
   b. II only
   c. neither I nor II

10. Which of the following statements is true?

    I: The worst-fit algorithm for bin packing always produces optimal results.

    II: The chromatic number of a graph is always less than the number of vertices of the graph.

    a. only I
    b. only II
    c. neither I nor II

CHAPTER 4

LINEAR PROGRAMMING

CHAPTER OBJECTIVES

Check off these skills once you feel you have mastered them.

- ❑ Create a chart to represent the given information in a linear programming problem with two variables.
- ❑ From its associated chart, write the constraints of a linear programming problem as linear inequalities.
- ❑ List two implied constraints in every linear programming problem.
- ❑ Formulate a profit equation for a linear programming problem when given the per-unit profits.
- ❑ Describe the graphical implications of the implied constraints listed in the third objective (above).
- ❑ Draw the graph of a line in a coordinate-axis system.
- ❑ Graph a linear inequality in a coordinate-axis system.
- ❑ Determine by a substitution process whether a point with given coordinates is contained in the graph of a linear inequality.
- ❑ Indicate the feasible region for a linear programming problem by shading the graphical intersection of its constraints.
- ❑ Locate the corner points of a feasible region from its graph.
- ❑ Evaluate the profit function at each corner point of a feasible region.
- ❑ Apply the corner point theorem to determine the maximum profit for a linear programming problem.
- ❑ Interpret the corner point producing the profit maximum as the solution to the corresponding linear programming problem.
- ❑ List two methods for solving linear programming problems with many variables.

GUIDED READING

Introduction

[text pp. 115–117]

key idea

Linear programming is used to make management decisions in a business or organizational environment. In this chapter, we are concerned with situations where resources—time, money, raw materials, labor—are limited. These resources must be allocated with a certain goal in mind: to do a job as quickly as possible, or to keep costs to a minimum, or to make a maximum profit. Such problems are modeled by systems of linear inequalities and solved by a combination of algebraic and numerical methods. We will focus on examples in the context of business, with the objective of making the largest possible profit.

Mixture Problems

[text pp. 117–118]

key idea

In a **mixture problem,** limited resources are combined to produce a product at maximum profit.

key idea

The two properties of an **optimal production policy** are: 1) it does not violate any of the limitations under which the manufacturer operates, such as the availability of resources; 2) it produces the maximum profit.

key idea

In a mixture problem, we seek an optimal production policy for allocating limited resources to make a maximum profit.

Mixture Problems Having One Resource

[text pp. 119–137]

key idea

A **production policy** is represented by a point in the **feasible region;** represented by a profit line.

question

Fred runs a small creamery which gets a shipment of 240 pints (30 gallons) of cream every day. When he produces ice cream, each container requires one-half pint of cream and sells at a profit of $.75. Determine the feasible region and profit formula for this example.

answer

Let x = number of containers of ice cream produced daily. Then the feasible region of x is:

$$0 \le x \le 480$$

The profit formula is:

$$P = .75x$$

explanation

Each container of ice cream uses .5 pint of cream, so x containers uses $.5x$ pints. Fred has only 240 pints, so $.5x \le 240$. Multiply by 2 and combine with the fact that x cannot be negative, so $x \ge 0$.

key idea

With several products and resources to consider, we can keep track of the quantities and conditions with a **mixture chart.**

key idea

The mixture chart has a row for each product and a column for each resource and for profit.

key idea

Fred's creamery adds sherbet as a new product. Each container of sherbet requires one-quarter pint of cream and sells at a profit of $.50. Construct a mixture chart for this situation.

answer

Let y = number of containers of sherbet produced daily. Here is the mixture chart:

	RESOURCES Pints of Cream = 240	PROFIT
Ice Cream (x units)	1/2	$.75
Sherbet (y units)	1/4	$.50

explanation

Note that resources are in columns and products are in rows.

key idea

The limits on a resource can be expressed as an inequality called a resource constraint.

question

Write a resource constraint for cream and the profit formula that now applies to Fred's creamery.

answer

The cream constraint is:

$$.5x + .25y \leq 240.$$

This can also be written as:

$$2x + y \leq 960$$

The profit formula is:

$$P = .75x + .50y$$

explanation

x containers of ice cream use $.5x$ pints of cream, y containers of sherbet use $.25y$ pints of cream. The two together cannot exceed the daily shipment of 240 pints. Multiply each term by 4 to get the second version. Similarly, the profit formula is the total profit from both ice cream and sherbet.

key idea

The resource constraints, together with the **minimum constraints,** can be used to draw a graph of the feasible region.

question

Fred has made agreements with his customers that obligate him to produce at least 100 containers of ice cream and 80 containers of sherbet. Draw a graph of the feasible region for Fred's creamery's production policy.

answer

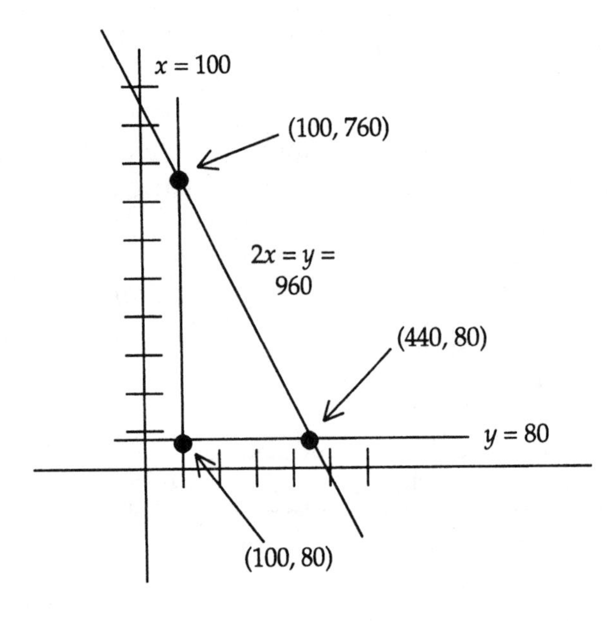

explanation

We add the conditions $x \geq 100$ and $y \geq 80$ to the resource inequality.

key idea

According to the **corner point principle,** the optimal production policy is represented by a **corner point** of the feasible region.

question

Find Fred's optimal production policy by evaluating the profit at each corner point of the feasible region.

answer

Fred's optimal policy is to produce 100 containers of ice cream and 760 containers of sherbet, yielding a profit of $455.

explanation

The profit values at the corner points are:

$$P\,(100, 760) = 455$$
$$P\,(440, 80) = 370$$
$$P\,(100, 80) + 115$$

So the chosen corner is the way to maximize profit.

question

Suppose a competitor drives down Fred's price of sherbet to the point that Fred only makes a profit of $.25 per container. What is Fred's optimal production policy?

answer

Fred's optimal policy would be to produce 440 containers of ice cream and 80 containers of sherbet, yielding a profit of $350.

explanation

The new profit formula would be:

$$P = .75x = .25y$$

question

What is Fred's new profit formula?

answer

$P = .75x + .25y$

explanation

The profit for ice cream remained unchanged at $.75x$. However, the profit for sherbet changed from $.50y$ to $.25y$. The profit values at the corner points would then be:

$$P\,(100, 760) = 265$$
$$P\,(440, 80) = 350$$
$$P\,(100, 80) = 95$$

Mixture Problems Having Two Resources

[text pp. 137–144]

key idea

With two resources to consider, there will be two resource constraints. The feasible region will generally be quadrilateral, with four corners to evaluate.

question

Fred's creamery decides to produce raspberry versions of both its ice cream and sherbet lines. Fred is limited to, at most, 600 pounds of raspberries, and he adds one pound of raspberries to each container of ice cream or sherbet. What is the new resource constraint for raspberries?

answer

The constraint for raspberries is:

$$x + y \leq 600$$

explanation

Each container of either product uses one pound of raspberries; $x + y$ is the total number of containers produced, and equals the number of pounds of raspberries.

question

Construct a mixture chart, taking into account both the cream and raspberry resource constraints.

answer

	RESOURCES		PROFIT
	Pints of Cream 240	Pounds of Raspberries 600	
Ice Cream (x units)	1/2	1	$.75
Sherbet (y units)	1/4	1	$.50

question

What does the feasible region look like for this new situation (with the same minimum constraints of 100 for ice cream and 80 for sherbet)?

answer

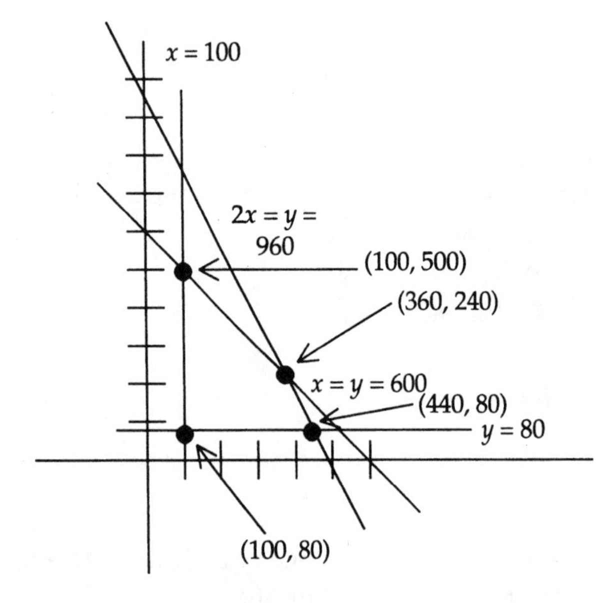

explanation

The raspberry constraint leads to the line $x + y = 600$. The intersection of this line with the line $2x + y = 960$ is the point (360, 240), a corner of the feasible region. Also, the point (100, 500) is now a corner.

question

What is Fred's optimal production policy (with the $.50 profit figure for sherbet)?

answer

Fred's optimal policy is to produce 360 containers of ice cream and 240 containers of sherbet, yielding a profit of $390.

explanation

The profit values at the corners are:

$$P\,(100, 500) = 325$$
$$P\,(360, 240) = 390$$
$$P\,(440, 80) = 370$$
$$P\,(100,80) = 115$$

so the chosen corner is the way to maximize profit.

The Corner Point Principle

[text pp. 144–149]

key idea

If you choose a theoretically possible profit value, the points of the feasible region yielding that level of profit lie along a profit line cutting through the region.

Profit Line: P = fixed value

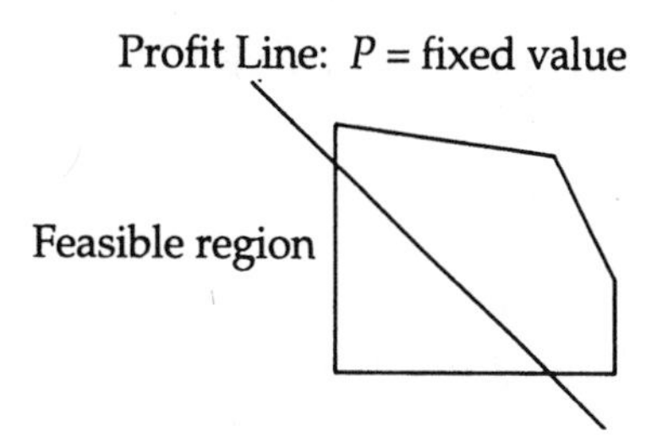

key idea

Raising the profit value generally moves the profit line across the feasible region until it just touches at a corner, which will be the point with maximum profit, the optimal policy.

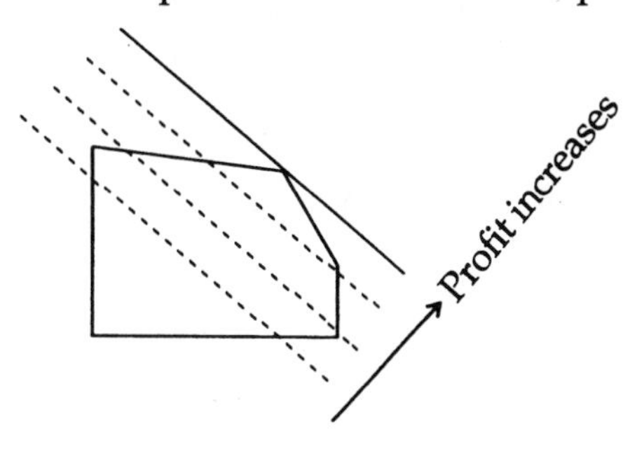

key idea

This explains the corner point principle; it works because the feasible region has no "holes" or "dents" or missing points along its boundary.

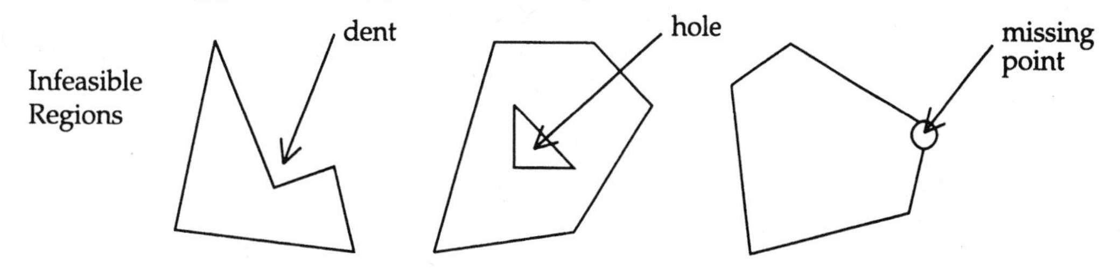

Linear Programming—The Wider Picture

[text pp. 149–154]

key idea

For realistic applications, the feasible region may have many variables ("products") and hundreds or thousands of corners. More sophisticated evaluation methods, such as the **simplex method,** must be used to find the optimal point.

key idea

The simplex method is the oldest algorithm for solving linear programming. It was devised by the American mathematician George Dantzig in the 1940's. Years later, Narendra Karmarker devised an even more efficient algorithm.

PRACTICE QUIZ

1. Where does the line $3x + 5y = 30$ cross the x-axis?
   a. at the point (10, 0)
   b. at the point (0, 6)
   c. at the point (3, 0)

2. Where do the lines $2x + 3y = 11$ and $y = 1$ intersect?
   a. at the point (1, 3)
   b. at the point (4, 1)
   c. at the point (0, 3 2/3)

3. Where do the lines $3x + 2y = 13$ and $4x + y = 14$ intersect?
   a. at the point (2, 3)
   b. at the point (1, 10)
   c. at the point (3, 2)

4. Which of these points lie in the region $3x + 5y \leq 30$?
   I. (6, 0)
   II. (1, 2)
   a. both I and II
   b. only II
   c. neither I nor II

5. What is the resource constraint for the following situation?
   Typing a letter (x) requires 4 minutes, and copying a memo (y) requires 3 minutes. A secretary has 15 minutes available.
   a. $x + y \leq 15$
   b. $4x + 3y \leq 15$
   c. $3x + 4y \leq 15$

6. What are the resource inequalities for the following situation?
   Producing a bench requires two boards and 10 screws. Producing a table requires 5 boards and 8 screws. Each bench yields \$10 profit and each table yields \$12 profit. There are 25 boards and 60 screws available.
   a. $2x + 10y \leq 25$
      $5x + 8y \leq 60$
      $x \geq 0, y \geq 0$
   b. $2x + 10y \leq 10$
      $5x + 8y \leq 12$
      $x \geq 0, y \geq 0$
   c. $2x + 5y \leq 25$
      $10x + 8y \leq 60$
      $x \geq 0, y \geq 0$

7. What is the profit formula for the following situation?
   Producing a bench requires two boards and 10 screws. Producing a table requires 5 boards and 8 screws. Each bench yields \$10 profit and each table yields \$12 profit. There are 25 boards and 60 screws available.
   a. $P = 25x + 60y$
   b. $P = 1/4x + 1/5\,y$
   c. $P = 10x + 12y$

8. Graph the feasible region identified by the inequalities:
   $3x + 2y = 13$
   $4x + y = 14$
   $x \geq 0, y \geq 0$

   Which of these points is in the feasible region?
   a. (2, 3)
   b. (0, 5)
   c. (4 1/3, 0)

9. Which of the following are true statements?
   I: The simplex algorithm always gives optimal solutions to linear programming problems.
   II: The optimal solution for a linear programming problem will always occur at a corner point.
   a. Only I is true.
   b. Only II is true.
   c. Both I and II are true.

10. Which of the following are true statements?
    I: The feasible region for a linear programming mixture problem never has holes in it.
    II: A linear programming mixture problem never involves more than two products.
    a. Only I is true.
    b. Only II is true.
    c. Both I and II are true.

CHAPTER 5

PRODUCING DATA

CHAPTER OBJECTIVES

Check off these skills once you feel you have mastered them.

- ❏ Identify the population in a given sampling or experimental situation.
- ❏ Identify the sample in a given sampling or experimental situation.
- ❏ Explain the difference between a population and a sample.
- ❏ Calculate the sample proportion given the sample size and number of positive responses.
- ❏ Determine from the sample proportion and the sample size the number of positive responses.
- ❏ Analyze a sampling example to detect sources of bias.
- ❏ Identify several examples of sampling that occur in our society.
- ❏ Select a numbering scheme for a population from which a random sample will be selected from a table of random digits.
- ❏ Use a table of random digits to select a random sample from a small population.
- ❏ Explain the difference between the experimental group and the control group in an experiment.
- ❏ Describe the placebo effect.
- ❏ Discuss why double blindness is desirable in an experiment.
- ❏ List two basic features of statistically designed experiments.
- ❏ Recognize the confounding of the effects of two variables in an experiment.
- ❏ Construct a Latin square to simplify the design of an experiment.
- ❏ Given a population proportion and its associated margin of error, list the range for a 95% confidence interval.

GUIDED READING

Introduction

[text p. 167]

key idea

Numbers are used in a myriad of ways to describe the world we live in. Our social and economic concerns, science and ecology, politics, religion, health, recreation, any area of human activity is better understood by the collection and analysis of data. It is vitally important to know how to produce trustworthy data, and how to draw reliable conclusions from them. This is the role of **statistics,** the science of data handling.

Sampling

[text pp. 168–169]

key idea

In statistical studies, we gather information about a small, partial group (a **sample**) in order to draw conclusions about the whole, large group we are interested in (the **population**).

question

In a study of the smoking habits of urban American adults, we ask 500 people their age, place of residence, and how many cigarettes they smoke daily. What is the population and what is the sample?

answer

The population is all Americans who live in a large city and are old enough to be classified as adults. The sample consists of those among the 500 interviewed who qualify as members of this population; for example, children would not be included, even if they smoke.

Bad Sampling Methods

[text pp. 169–171]

key idea

Sampling a population by selecting members easiest to reach produces a **convenience sample,** which often leads to unrepresentative data.

key idea

Systematic error caused by bad sampling methods may lead to a **biased** study favoring certain outcomes.

question

Customers at a supermarket are sampled to determine their opinion about a volatile political issue. Can you identify a possible source of bias in such a survey?

answer

There may be gender-based bias, with women overrepresented in the sample. Depending on the location of the market, there may also be a bias according to economic class, education level, political affiliation, etc.

key idea

A sample of people who choose to respond to a general appeal is called a **voluntary response sample.** This voluntary response is a likely source of bias.

question

Television viewers are invited to call an 800 number to report their opposition to a bill to increase state gasoline taxes. Why might this survey be biased?

✲ **answer**

There is a high likelihood that a disproportionately large number of people angry about a potential tax increase will take the trouble to call to register their opposition.

Simple Random Samples

[text pp. 171–175]

key idea

We can use a **simple random sample** (SRS) to eliminate bias. This is the equivalent of choosing names from a hat; each individual has an equal chance to be selected.

question

To choose a sample of five cards from a deck of 52 cards, you shuffle the cards and choose the first, third, fifth, seventh, and ninth card. Will this lead to a simple random sample?

✲ **answer**

Yes, if the deck has been thoroughly shuffled.

explanation

After shuffling, any given card is equally likely to occupy any given position in the deck.

key idea

A two-step procedure for forming a SRS using a **table of random digits** is:

Step 1. Give each member a numerical label of the same length.

Step 2. Read from the table strings of digits of the same length as the labels. Ignore groups not used as labels and also ignore any repeated labels.

question

Describe how to use the table of random digits to form a random sample of 75 students at Hypothetical University from the entire population of 1350 HU students.

✲ **answer**

Assign each HU student a four digit numerical label, 0001–1350, making sure that no label is assigned twice. Then starting anywhere in the random digit table, read strings of 4 consecutive digits, ignoring repetitions and unassigned strings, until 75 assigned labels are obtained. The students with those labels are the sample.

key idea

Multistage random sampling has some practical advantages over the SRS method, but is not quite as accurate and the results are more difficult to analyze. However, for practical reasons this type of sampling is frequently used, particulary in national polls and surveys.

Statistical Estimation

[text pp. 175–179]

key idea

A sample should resemble the population, so that a **sample statistic** can be used to estimate a characteristic of the population.

question

A random sample of 150 people are asked if they own dogs, and 57 of them say yes. What would you estimate the percentage of dog owners to be in the general population?

answer

The best estimate is 38%.

explanation

The sample proportion is 57/150 = .38 = 38%; the actual population proportion may differ somewhat, but is reasonably likely to be fairly close to that of the sample.

key idea

Results of a survey will vary from sample to sample; this is called sampling variability. Data from repeated sampling can be recorded in a **histogram,** a bar graph representing frequencies.

question

Suppose the previous dog survey was conducted simultaneously by twelve investigators, each sampling 150 people, leading to the following twelve percentages of dog owners: {39%, 37%, 37%, 39%, 40%, 38%, 41%, 40%, 39%, 41%, 42%, 39%}. Sketch a histogram for this data.

answer

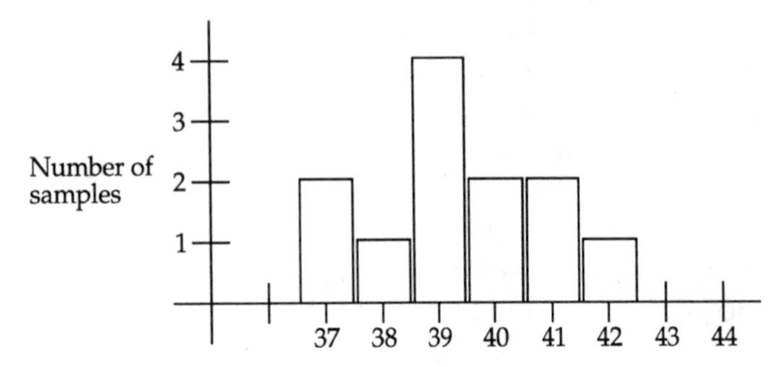

explanation

Two of the samples had a proportion of 37%, so the bar over the 37% marker has height 2, and so on.

key idea

A **sampling histogram** will generally display a regular pattern with two important features:
The results will be centered symmetrically around a peak, the true population value.
The spread of the data will be tighter for large sample sizes, wider for small ones.

key idea

The **margin of error** of a survey gives an interval that includes 95% of the samples and is centered around the true population value.

question

Suppose the results of the dog survey are announced as follows: "The percentage of people who own dogs is 39%, with a margin of error of 4%". Can you say the following with reasonable (about 95%) certainty of being right?

(a) At least (that is, not less than) 39% of people own dogs.
(b) At most (that is, not more than) 45% of people own dogs.

answer

(a) No
(b) Yes

explanation

(a) The true percentage of dog owners is just as likely to be below 39% as above it.
(b) With a 4% margin of error, the true percentage is almost certain to be within the 35%–43% range, and is thus highly likely to be less than 45%.

Experiments

[text pp. 179–180]

key idea

An **observational study** is a passive study of a variable of interest; an **experiment** is an active trial of a treatment and its effects.

question

Which is an experiment and which is an observational study?

(a) You ask a sample of smokers how many cigarettes they smoke daily, and measure their blood pressure.
(b) You select a sample of smokers and measure their blood pressure. Then you ask them to reduce their smoking by 5 cigarettes a day; after 3 months you recheck their blood pressure.

answer

(a) is an observational study, (b) is an experiment.

explanation

In (b), we are actively influencing the behavior of the subjects, whereas in (a), we are passively observing and measuring.

key idea

When designing an uncontrolled study, care must be taken to avoid **confounded variables** —variables whose effects on the outcome cannot be distinguished from one another.

Randomized Comparative Experiments

[text pp. 181–185]

key idea

We can reduce the effect of confounded variables by conducting a **randomized comparative experiment.** The sample for the experiment is matched by a **control group,** with subjects assigned randomly to the treatment or the control group.

question

How would you design a simple randomized comparative experiment to test the effect of a high-potassium diet on smokers' blood pressure? Assume you have 200 smokers who have agreed to participate in the experiment.

answer

From the group of 200 smokers who have agreed to participate, randomly select 100 to try the high-potassium diet. The other 100 will serve as the control group, and will make no change in their diet. Measure the blood pressure of each subject at the beginning and end of the testing period, and compare changes in the two groups.

key idea

To avoid the **placebo effect** and any possible bias on the part of the experimenters, use a **double-blind experiment,** so neither subjects nor investigators know which treatment an individual is receiving.

question

How would you design a double-blind experiment to test the effect of a vitamin supplement on smokers' blood pressure?

answer

Randomly assign labels 1–200 to your subjects. Using the labels, randomly choose 100 of the subjects to receive the vitamin supplement (say, in pill form), while the other group receives an indistinguishable placebo. The list of which group each subject belongs to is kept confidential; neither subjects nor experimenters know who is taking the real supplement until the experiment is over and the data have been recorded.

explanation

This way, neither psychological factors nor unconscious bias on the part of the experimenters can play a role.

key idea

In order to perform a **two-factor experiment,** four sample groups are needed.

question

Describe the subject groups for a combined test of a high-potassium supplement and a multi-vitamin supplement on smokers' blood pressure.

answer

Form four equal groups using a random procedure. Give the groups treatments according to the following table:

Group I:	High-potassium supplement Multi-vitamin supplement	Group II:	High-potassium supplement Placebo
Group III:	Placebo Multi-vitamin supplement	Group IV:	Placebo Placebo

The experiment should be double-blind, so no participant knows who belongs to which group.

Statistical Evidence

[text pp. 185–187]

key idea

Small differences between groups in a study can be due to random variation, but **statistically significant** differences are too large to be attributable to chance and are reliable evidence of a real effect of the factors being studied.

key idea

A **prospective study** is an observational study that records slowly developing effects of a group of subjects over a long period of time.

key idea

Only **experimentation** can produce fully convincing statistical evidence of cause and effect.

key idea

A well-designed experiment is one that uses the principles of comparison and randomization: comparison of several treatments and the random assignment of subjects to treatments. If subjects are randomly assigned to treatments, we can be confident that any differences among treatment groups that are too large to have occurred by chance are statistically significant.

Statistics in Practice

[text pp. 187–189]

key idea

Even a sound statisitical design cannot guard against some of the pitfalls associated with statisitical experiments. For example, nonresponse can be a cause of bias in an experiment, as can the artificial environments created for some experiments.

PRACTICE QUIZ

1. A marketing firm interviewed 80 shoppers randomly selected from the 4000 customers at one of the mall's 45 stores yesterday. The sample in this situation is the
   a. 45 stores.
   b. 80 selected shoppers.
   c. 4000 customers.

2. In an election for mayor, there are 3 candidates and 24,000 eligible voters. A newspaper interviews 240 voters as they come out of the polls. The population here is the
   a. 240 voters interviewed.
   b. 3 candidates.
   c. 24,000 eligible voters.

3. A well-designed survey should minimize
   a. bias.
   b. randomness.
   c. the placebo effect.

4. To determine the food preferences of students, a staff member surveys students as they exit a local bar. This type of sample is a
   a. convenience sample.
   b. voluntary response sample.
   c. simple random sample.

5. A survey on the benefits of jogging is conducted outside a sporting-goods store. This is an example of
   a. bias.
   b. confounding.
   c. the placebo effect.

6. In a district with 37,500 voters, 1500 are sampled and 1200 prefer your candidate. The sample percent is
   a. 80%.
   b. 37.5%.
   c. 4%.

7. Here is a list of random numbers: 16807 64853 17463 14715. Use this list to choose 3 numbers from the set {1, 2, 3, . . . , 20}. What are the numbers?
   a. 16, 7, 17
   b. 16, 17, 15
   c. 16, 17, 14

8. If Adam's sample statistic has a margin of error of 3% and Ben's sample statistic has a margin of error of 5%, then
   a. Ben's estimate is biased.
   b. Adam has more samples.
   c. Ben's experiment gave a higher sample estimate.

9. Five workers each sample 50 students to determine their favorite fast-food restaurant. Each worker returns with slightly different results. This is probably due to
   a. bias.
   b. sampling variability.
   c. the use of a control group.

10. A drug test randomly selects one of three treatments for each participant. Neither the experimenter nor the participant knows which drug is chosen. This is an example of
    a. a randomized comparative experiment.
    b. an experiment which is not double-blind.
    c. bad sampling methods.

CHAPTER 6

EXPLORING DATA

CHAPTER OBJECTIVES

Check off these skills once you feel you have mastered them.

- ❑ Calculate the mean of a set of data.
- ❑ After sorting a data set, determine its median.
- ❑ Determine both lower and upper quartiles for a given data set.
- ❑ Calculate the range of a given data set after observing its largest and smallest individual observations.
- ❑ List the five-number summary for a given data set.
- ❑ Construct the diagram of a boxplot when given the data set's five-number summary.
- ❑ Construct a histogram for a small data set.
- ❑ Identify from a histogram possible outliers of a data set.
- ❑ List and describe two types of distributions for a histogram.
- ❑ Draw a scatterplot for a small data set consisting of pairs of numbers.
- ❑ From a scatterplot, draw an estimated line fit.
- ❑ Describe how the concept of distance is used in determining a least-squares regression line.
- ❑ Use the given equation of a regression line to predict response (Y) values from given causal (X) values.
- ❑ Calculate the correlation between two quantitative variables, one explanatory and one response, from a data set.
- ❑ Understand the significance of the correlation between two variables, and estimate it from a scatterplot.

GUIDED READING

Introduction

[text p. 204]

Data, or numerical facts, are essential for making decisions in almost every area of our lives. But to use them for our purposes, huge collections of data must be organized and distilled into a few comprehensible summary numbers and visual images. This will clarify the results of our study, and allow us to draw reasonable conclusions. The analysis and display of data are thus the groundwork for statistical inference.

Exploratory data analysis combines numerical summaries with graphical display to see patterns in a set of data.

The organizing principles of data analysis are:

1) Examine individual variables, and then look for relationships between several variables.
2) Draw a graph and add to it numerical summaries.
3) Look first for an overall pattern and then for significant deviations from the pattern.

Displaying Distributions

[text pp. 205–208]

key idea

We can visualize and understand the **distribution** of a variable (that is, the **frequencies** of its values) by the use of a **dotplot** display.

question

The scores of 20 students on a Chemistry 101 quiz are as follows:

8 5 7 8 4 3 7 6 7 6 5 8 9 4 6 9 0 5 10 7

Display the distribution with a dotplot.

answer

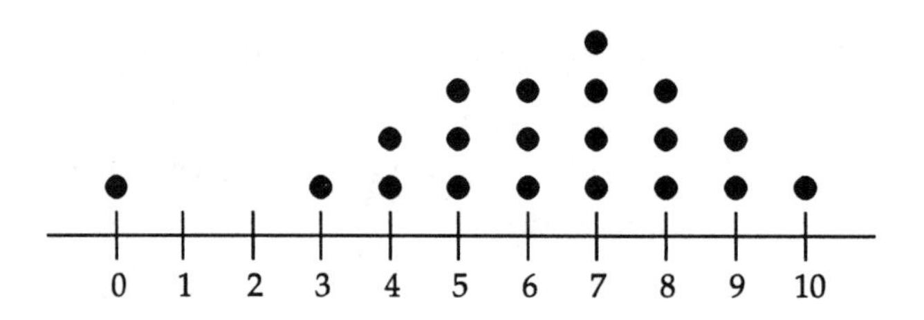

explanation

The data contains a single 0, a single 3, two 4's, three 5's, and so on.

key idea

Unusual circumstances may be reflected by the presence of an **outlier** in the data.

question

Find an outlier in the chemistry quiz data and give a possible explanatory circumstance.

answer

The score of 0 is clearly an outlier.

explanation

With a 0 on a quiz, the most likely explanation is that the student missed the quiz. It is also possible that the student was completely unprepared, and made poor guesses.

key idea

When the data set is very large or does not have well-separated values, a **histogram** is more appropriate than a dotplot for visual display. A histogram shows the **frequencies** of data within each range.

question

In the same chemistry course, the professor converted quiz averages to final grades according to these ranges:

A	B+	B	C+	C	D	F
89–100	85–88.9	80–84.9	75–79.9	70–74.9	65–69.9	<65

The quiz averages of the 20 students were:

72 85.5 93.5 68 73.5 82.5 80 79.5 56.5 87.5 89.5 71 79.5 86 75 76.5 83 86.5 78 67

Draw a histogram of the grade distribution.

answer

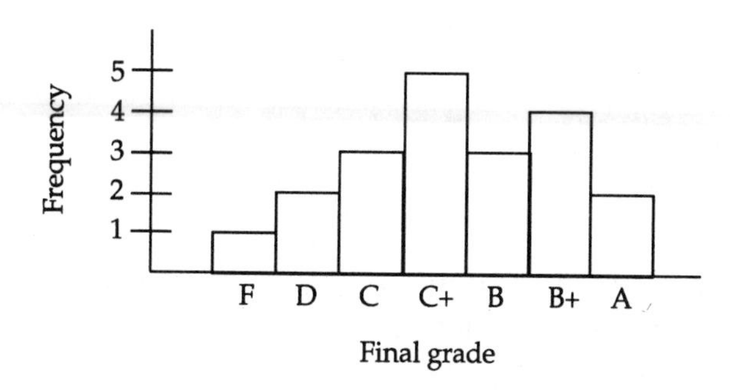

explanation

The height of the bar over a grade tells you how many students earned that grade. For example, there were 4 students with averages in the 85–88.9 range, so the B+ bar has height 4.

Interpreting Histograms

[text pp. 208–210]

key idea

The important features of a **histogram** are its overall shape, its center, and its spread. We may also observe outliers.

key idea

A distribution may be **symmetric,** or it may be **skewed** to the left or to the right.

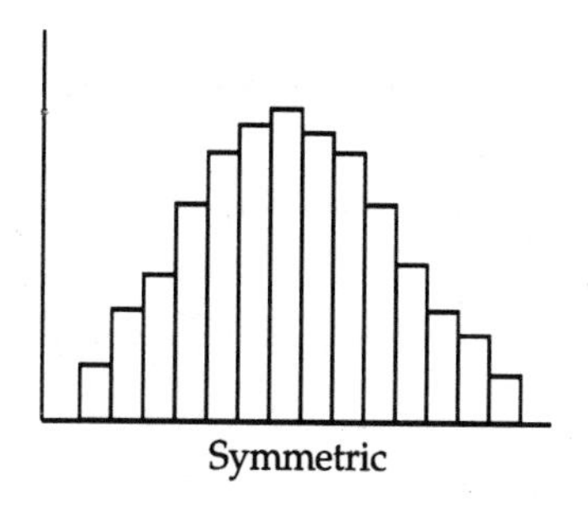
Symmetric

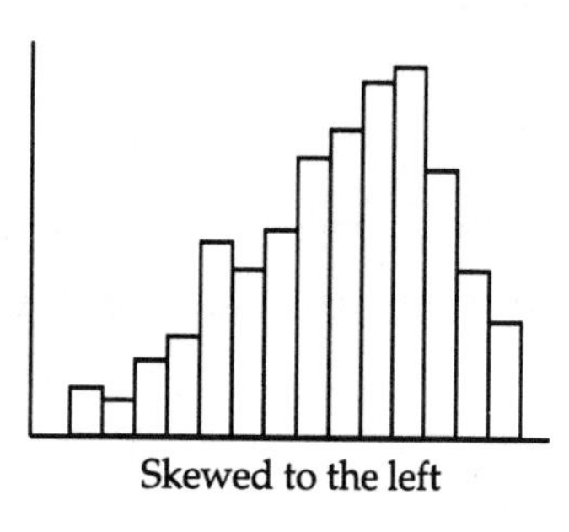
Skewed to the left

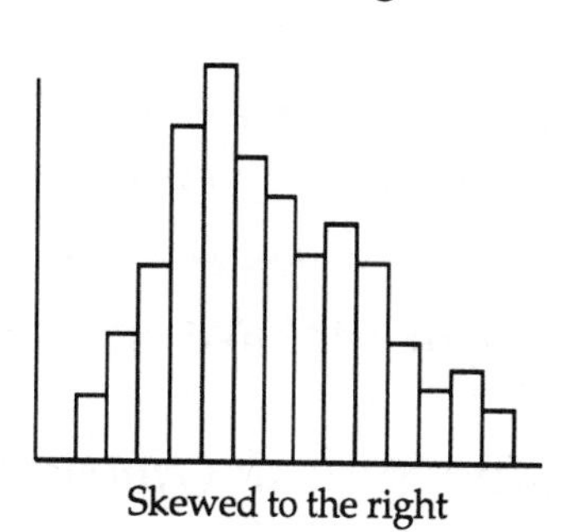
Skewed to the right

Describing Center: Mean and Median

[text pp. 213–215]

key idea

The **mean** of a data set is obtained by adding the values of the observations in the data set and dividing by the number of data. If the observations are listed as values of a variable $\bar{x}$, then the mean is written as x.

question

Calculate the means of the previous data sets (shown again below).

(a) 9 1 3 6 4 5 5 4 7
(b) 7 2 3 4 5 3 1 7 2 9 3 6 5 8 6 3

answer

(a) $\bar{x} = 4.89$
(b) $\bar{x} = 4.625$

explanation

(a) The sum of the numbers is 44, and there are 9 numbers; 44 / 9 = 4.89 (approx.).
(b) The sum of the numbers is 74, and there are 16 numbers; 74 / 16 = 4.625.

key idea

The **median** of a distribution is a number in the middle of the data, so that half of the data is above the median, and the other half is below it.

question

What are the medians of the data sets below?

(a) 9 1 3 6 4 5 5 4 7
(b) 7 2 3 4 5 3 1 7 2 9 3 6 5 8 6 3

answer

(a) 5 (b) 4.5

explanation

(a) Arrange the data in order:

1 3 4 4 **5** 5 6 7 9

The middle number is 5.

(b) Arrange the data in order:

1 2 2 3 3 3 3 **4** **5** 5 6 6 7 7 8 9

The median is the average of the two middle numbers, 4 and 5.

Describing Spread: Quartiles

[text pp. 215–217]

key idea

The **quartiles** Q_1 (the point below which 25% of the observations lie) and Q_3 (the point below which 75% of the observations lie) give a better indication of the true spread of the data.

question

What are the quartiles of the previous data set (shown again below)?

7 4 10 8 5 6 4 6 1 3 7 5

answer

$Q_1 = 4$ and $Q_3 = 7$.

explanation

Arrange the data in order and in 4 equal blocks:

1 3 **4** **4** 5 5 6 6 **7** **7** 8 10

The median of the lower half of the data is 4, and the median of the upper half of the data is 7.

The Five-Number Summary and Boxplots

[text pp. 217–218]

key idea

The **range,** the difference between the highest and the lowest observations of a data set, depends only on the most extreme data points.

question

What is the range of this data set?

7 4 10 8 5 6 4 6 1 3 7 5

answer

The range is 9.

explanation

The extreme values are 1 (low) and 10 (high), and $10 - 1 = 9$.

key idea

The **five-number summary** consists of the median, quartiles, and extremes (high and low), and can be visualized with a special graph called a **boxplot.**

question

List the five-number summary of the previous data set (shown again below) and draw a boxplot.

7 4 10 8 5 6 4 6 1 3 7 5

answer

The five-number summary is:

1 4 5.5 7 10

The boxplot (drawn horizontally) looks like this:

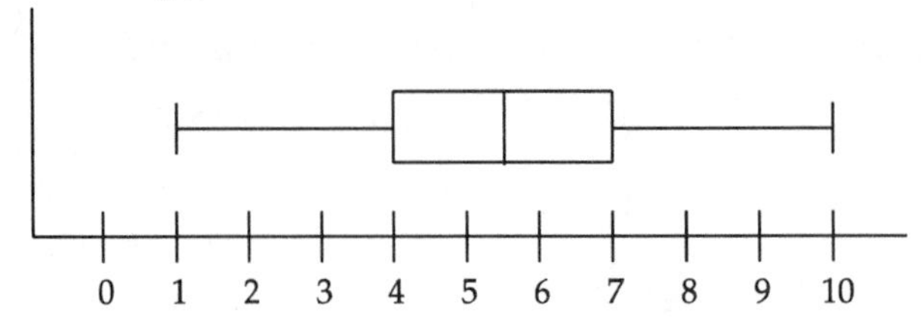

explanation

The data set has median 5.5, quartiles 4 and 7, and extremes 1 and 10.

Describing Spread: The Standard Deviation

[text pp. 218–221]

key idea

The **variance** (s^2) of a set of observations is an average of the squared differences between the individual observations and their mean value.

question

Find the variance of the data set below:

2 4 8 3 5 6 2 6 1 3

answer

$s^2 = 4.89$

explanation

The mean of the data is

$$\bar{x} = (2 + 4 + 8 + 3 + 5 + 6 + 2 + 6 + 1 + 3) / 10 = 40 / 10 = 4$$

The table for calculating the variance s^2 is

Values of x	2	4	8	3	5	6	2	6	1	3	
Deviations of $x - \bar{x}$	–2	0	4	–1	1	2	–2	2	–3	–1	
Squared deviations	4	0	16	1	1	4	4	4	9	1	Sum = 44

$s^2 = 44 / 10 - 1 = 4.89$
Thus $s^2 = 44 / 9 = 4.89$.

key idea

The **standard deviation** (s) is the square root of the variance and measures the spread of the data around the mean in the same units of measurement as the original data set.

question

Find the standard deviation of the previous data set (shown again below).

2 4 8 3 5 6 2 6 1 3

answer

$s = 2.21$

explanation

The variance is 4.89; taking the square root yields 2.21.

Displaying Relations Between Two Variables

[text pp. 221–223]

key idea

Graphs are useful for recognizing connections between two variables. A **scatterplot** is the simplest such representation, showing the relationship between an **explanatory variable** (on the horizontal axis) and a **response variable** (on the vertical axis).

key idea

We look for an overall straight-line pattern in the scatterplot. A stronger relationship would yield points quite close to the line, a weaker one would have more points scattered around the line.

question

Draw a scatterplot showing the relationship between the observed variables x and y, with the data given in the table below.

Value of x	2.2	3.5	7.1	5.4	1.6	4.5	6.2	3.8	4.9	6.8
Value of y	1.0	1.5	4.0	2.3	1.1	2.0	3.9	2.0	2.8	2.8

answer

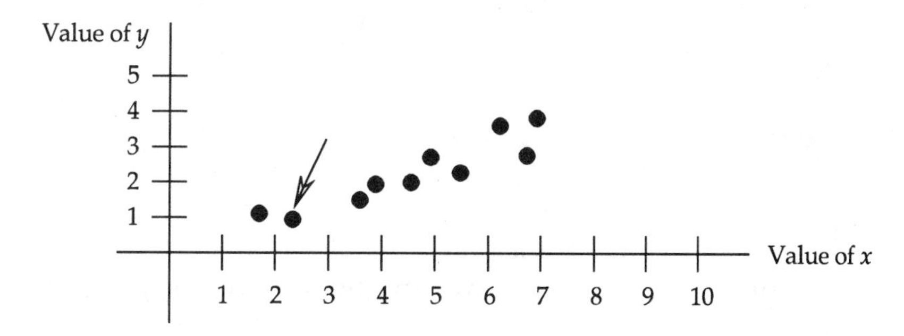

explanation

Each pair of observations (x, y) yields a point in the scatterplot. For example, the first pair (2.2, 1.0) yields the point indicated by an arrow.

Regression Lines

[text pp. 223–225]

key idea

A straight line drawn through the heart of the data and representing a trend is called a **regression line**, and can be used to predict values of the response variable.

key idea

The equation of a regression line will be $y = a + bx$, where a is the intercept and b is the slope of the line.

key idea

Starting with the scatterplot of the data from the previous section, we can draw a line angling up through the data:

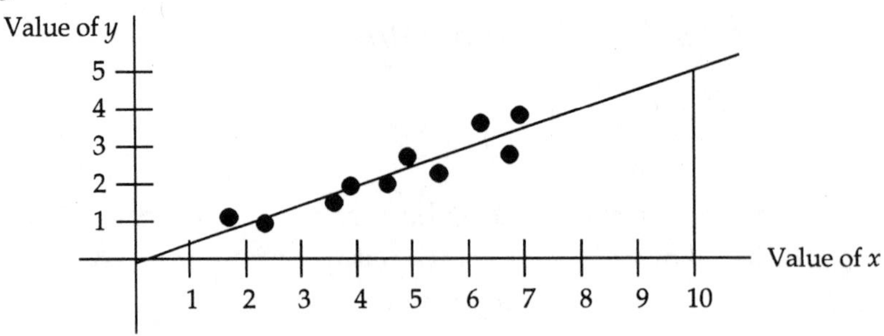

The equation of this regression line is $y = -.0238 + .5139x$ (this was obtained from a computer program).

question

Use the regression line to predict a value for y if x takes the value 8.5.

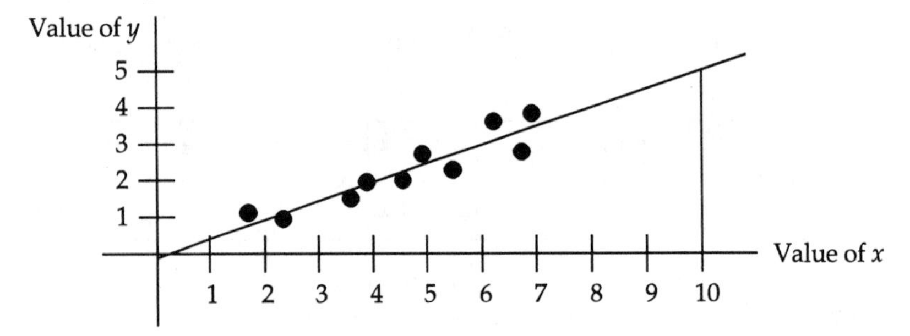

answer

An approximate value is $y = 4.3$.

explanation

We can use the scatterplot, with a theoretical point that has an x value of 8.5, and then read across to find the appropriate y value (as shown by the dotted lines).

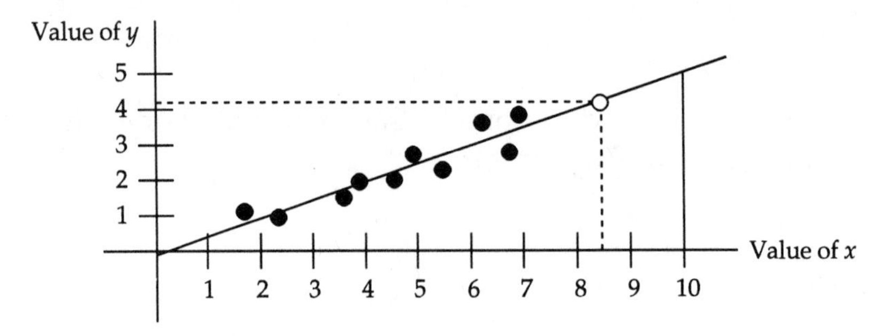

Alternatively, we can use the regression line equation $y = -0238 + .5139x$ and substitute $x = 8.5$, to calculate $y = 4.3443$, or 4.3.

Correlation

[text pp. 225–228]

key idea

The **correlation** r measures the strength of the linear relationship between two quantitative variables; r always lies between –1 and 1.

key idea

Positive r means the quantities tend to increase or decrease together; negative r means they tend to change in opposite directions, one going up while the other goes down. If r is close to 0, that means the quantities are fairly independent of each other.

question

Give a rough estimate of the correlation between the variables in each of these scatterplots:

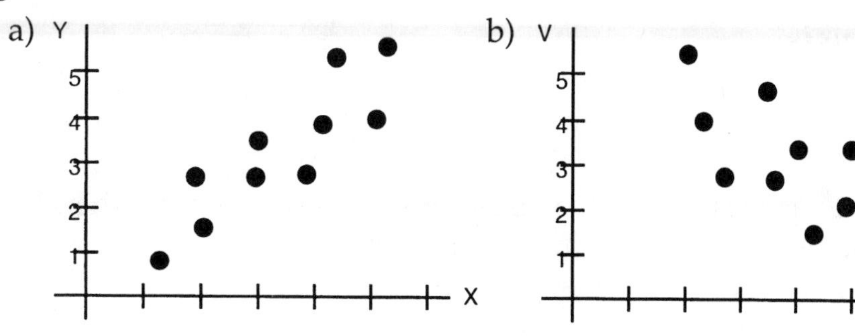

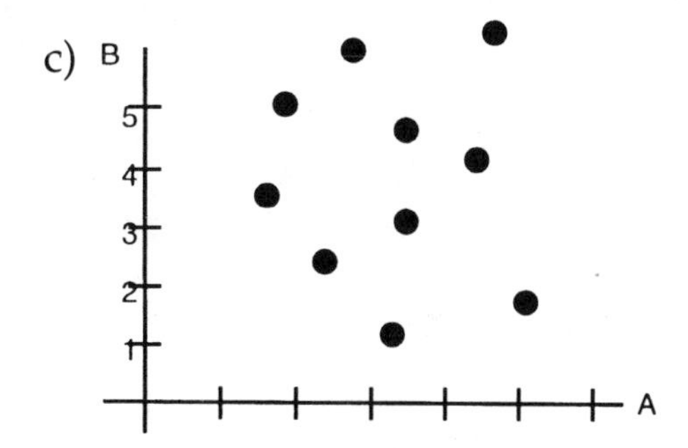

answer

a) $r \approx 1$ b) $r \approx -1$ c) $r \approx 0$

explanation

a) The points have a fairly tight linear relationship with a positive association, the variables X and Y increasing and decreasing together.
b) The points have a strong negative association, with high values of U associated with low values of V, and vice versa.
c) The variables A and B are fluctuating independently, with no clear correlated trend.

key terms and phrases

- correlation

Least-Squares Regression

[text pp. 228–230]

key idea

The **least-squares regression line** runs through a scatterplot of data so as to be the line that makes the sum of the squares of the vertical deviations from the data points to the line as small as possible. This is often thought of as the "line of best fit" to the data.

key idea

There is a formula for the equation of the least-square regression line for a data set on an explanatory variable x and a response variable y. The formula depends on knowing the means of x and y, the standard deviations of x and y and their correlation r. It produces the **slope** and **intercept** of the regression line.

Modern Data Analysis

[text pp. 230–232]

key idea

Sophisticated statistical analysis software packages use "scatterplot smoothers" to detect and display complex non-linear patterns in data as lines.

key idea

Looking at data using several methods (scatterplots, boxplots, 3-D graphs using computer graphics) often yields results that we wouldn't see using just one method.

key terms and phrases

- least-squares regression line
- slope
- intercept

PRACTICE QUIZ

1. The weights (in pounds) of your cousins are: 120, 89, 108, 76, 21. Which are the outliers?
   a. 21 only
   b. 120 only
   c. both 120 and 21

2. Below is a stemplot of the ages of adults on your block. Which statement is true?

```
2 | 1125
3 | 025788
4 | 15
5 | 257
6 | 25
7 | 8
8 | 1
```

   a. The stemplot is roughly symmetric.
   b. The stemplot is skewed toward the lower ages.
   c. The two oldest people are outliers.

3. Here are 7 measured lengths (in inches): 13, 8, 5, 3, 8, 9, 12. Find their median.
   a. 3
   b. 8
   c. 8.3

4. Here are 7 measured lengths (in inches): 13, 8, 5, 3, 8, 9, 12. Find their mean.
   a. 3
   b. 8
   c. 8.3

5. The boxplot graph always includes the
   a. mean and median.
   b. quartiles and the standard deviation.
   c. quartiles and the median.

6. The percentage of scores on a standardized exam that lie between the first and third quartiles is
   a. 25%.
   b. 50%.
   c. 75%.

7. If the mean of the data 2, 4, 6, 3, 5, 8, 7 is 5, what is its standard deviation?
   a. $\sqrt{\frac{12}{7}}$
   b. 4
   c. 2

8. The park wants to predict the amount of ice used each day, based on the predicted high temperature. Which variable would be the explanatory variable?
   a. the amount of ice used
   b. the predicted high temperature
   c. the actual high temperature

9. The daily ice consumption (in pounds) y at a park is related to the predicted high temperature (in degrees F) x. Suppose the least-squares regression line is $y = 250 + 25x$. Today's predicted high temperature is 90 degrees F. This means that
   a. at least 2500 pounds of ice will be needed today.
   b. approximately 2500 pounds of ice will be needed today.
   c. exactly 2500 pounds of ice will be needed today.

10. The daily ice consumption (in pounds) y at a park is related to the predicted high temperature (in degrees F) x. Suppose the least-squares regression line is $y = 250 + 25x$. Suppose 2300 pounds of ice were used yesterday. This leads you to believe that yesterday's high temperature was closest to
    a. 52 degrees F
    b. 82 degrees F
    c. 92 degrees F

CHAPTER 7

PROBABILITY: THE MATHEMATICS OF CHANCE

CHAPTER OBJECTIVES

Check off these skills once you feel you have mastered them.

- ❑ Describe the sample space for a given random phenomenon.
- ❑ Explain what is meant by the probability of an outcome.
- ❑ List two laws of probability.
- ❑ Apply the laws of probability to determine the validity of a probability model.
- ❑ Identify which probability law is not satisfied for a given illegitimate probability model.
- ❑ Construct a tree diagram to analyze a simple binomial probability example.
- ❑ Given a tree diagram, list the sample space and assign probabilities for the outcomes so that the laws of probability are satisfied.
- ❑ Compute the probability of an event when the probability model of the experiment is given.
- ❑ Describe a normal curve.
- ❑ Locate the mean and standard deviation from a graph of a normal curve.
- ❑ Explain the 68–95–99.7 rule.
- ❑ Apply the 68–95–99.7 rule to compute normal probabilities.
- ❑ Given the mean and standard deviation of a normally distributed data set, compute the percent of the population that falls within a given interval.
- ❑ Compute the expected value of an outcome when the associated probability model is defined.
- ❑ Explain the significance of the central limit theorem.
- ❑ Apply the addition rule to calculate the probability of a combination of several disjoint events.
- ❑ Draw the probability histogram of a probability model, and use it to determine probabilities of events.

GUIDED READING

Introduction

[text pp. 251–252]

Like a roll of the dice or a coin flip, a repeatable phenomenon is **random** if any particular outcome is quite unpredictable, while in the long run, after a large number of repeated trials, a regular, predictable pattern emerges. Of course, games of chance are an obvious application of the laws of randomness, but much more fundamental areas of human and natural activity are subject to these laws. Physics, genetics, economics, politics, and essentially any area in which large numbers of people or objects are examined or measured, can best be understood via the mathematics of chance.

What is Probability?

[text pp. 252–253]

key idea

Any specific outcome of a random phenomenon will tend to occur with predictable frequency over a very long sequence of trials. This frequency is called the **probability** of the outcome.

key idea

If you tossed a coin 10 times and got 7 heads, you would not be at all surprised, because even though it represents a frequency of 70%, the number of trials is too small to draw any conclusion.

question

If you tossed a coin 1000 times and got 507 heads, what would you think of the coin?

answer

The coin is probably an evenly-weighted coin.

explanation

The observed frequency is 507/1000, or approximately 50%, which is the correct probability of heads with a normal coin.

question

If you tossed a coin 1000 times and got 711 heads, what would you think of the coin?

answer

The coin is probably not a normal coin, but one strongly weighted towards heads.

explanation

After a long series of trials, the observed frequency of 711/1000, or approximately 70%, is a large deviation from the normal 50% probability.

Probability Models

[text pp. 253–255]

key idea

We begin the description of a random phenomenon by listing the **sample space** (S) of all possible outcomes.

key idea

For example, the sample space for tossing a coin is {H, T}. The sample space for rolling a single ordinary die is {1, 2, 3, 4, 5, 6}.

question

What is the sample space for rolling two dice and recording the *sum* of the up faces?

answer

{2, 3, 4, 5, 6, 7, 8, 9, 10, 11, 12}

explanation

When you add up two numbers between 1 and 6, you get all the numbers between 2 and 12 (inclusive) as possible sums.

key idea

A **probability model** is a mathematical description of a random phenomenon given by a sample space S, along with assigned probabilities $P(s)$ for each outcome s in S. These probability numbers represent the expected proportions or frequencies of occurrence of the outcomes.

question

Complete these rules:

(a) Each probability $P(s)$ is a number between ________________.

(b) All the probabilities $P(s)$ add up to ___________.

answer

(a) Each probability $P(s)$ is a number between 0 and 1.
(b) All the probabilities $P(s)$ add up to 1.

explanation

Each outcome occurs with some frequency between never (0) and always (1), and 100% of the time one of the possible outcomes occurs.

key idea

The probability of a collection of outcomes (an **event**) is the sum of the probabilities of the outcomes that constitute the event.

question

Write the probability model for rolling one ordinary die, and calculate the probability for the event that the up face is an odd number.

answer

The probability model is:

Up face	1	2	3	4	5	6
Probability	1/6	1/6	1/6	1/6	1/6	1/6

The probability of the up face being an odd number is $P(\text{odd}) = 3/6 = 1/2 = .5 = 50\%$.

explanation

The odd outcomes are 1, 3, and 5. Each outcome occurs a sixth of the time, so we calculate $1/6 + 1/6 + 1/6 + 3/6 + 1/2$.

Probability Rules

[text pp. 255–259]

key idea

The **addition rule** allows us to add up the probabilities of two or more **disjoint events.** Since no two outcomes are held in common by any pair of the events, there is no double counting of any outcomes when we take the sum.

key idea

The **probability histogram** of a probability model shows graphically the likelihood of each outcome. For example, here is the histogram of the standard model for rolling two dice and adding up the two results:

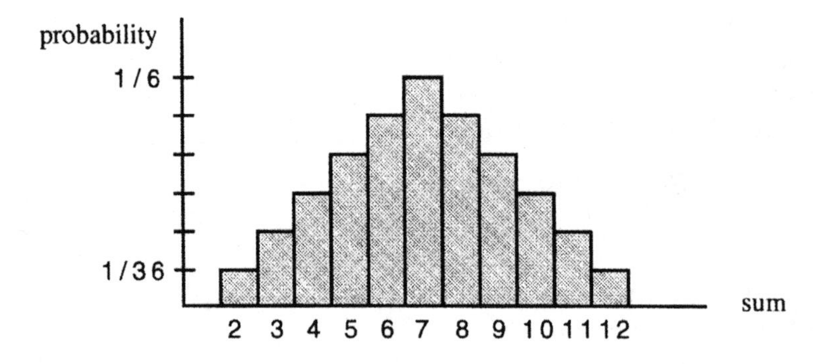

question

If you roll two dice, what is the probability of rolling a total less than or equal to 6?

answer

15/36

explanation

A total of 6 or less means one of these events: {2, 3, 4, 5, 6}. The addition rule says to add up their probabilities, and these can be read from the histogram:

$$p(2) = 1/36, p(3) = 2/36, p(4) = 3/36, p(5) = 4/36, p(6) = 6/36.$$

Thus $p(\text{total}^2\ 5) = 1/36 + 2/36 + 3/36 + 4/36 + 5/36 = 15/36$

key terms and phrases

- disjoint events
- addition rule for disjoint events
- probability histogram

Equally Likely Outcomes

[text pp. 259–262]

key idea

In many important probability models, all outcomes are equally likely to occur, so that all the probabilities must be equal. If there are n outcomes, each outcome has $P = 1/n$. If an event E consists of k outcomes, then $P(E) = k/n$.

For example, when rolling two ordinary dice, there are 36 possible outcomes: (1, 1), (1, 2), and so on up to (6, 6). Each of these is equally likely, so each has the probability $P = 1/36$.

question

When rolling two dice, what is the probability of obtaining a sum of 10?

answer

1/12

explanation

There are three combinations of two dice that will add to 10: (4, 6), (5, 5), (6, 4). Thus, the probability is $3/36 = 1/12$.

key idea

With equally likely outcomes, probability calculations come from **combinatorics,** or the study of counting methods.

(a) Arranging k objects chosen from a set of n possibilities, with repetitions allowed, can be done in n^k distinct ways.

(b) Arranging k objects chosen from a set of n possibilities, with repetitions allowed, can be done in $nx(n - 1)x \ldots x(n - k = 1)$ distinct ways (notice there are k factors here).

question

(a) How many code words (that is, strings of letters) of length four can be formed that use only the vowels {A, E, I, O, U, Y}?

(b) How many of these words have no letter occurring more than once?

answer

(a) 1296

(b) 360

explanation

(a) $n = 6$ and $k = 4$, use principle A to obtain $6^4 = 1296$.

(b) Same n and k, but use principle (B) to get $6 \times 5 \times 4 \times 3 = 360$.

The Mean of a Probability Model

[text pp. 262–265]

key idea

In using probability to help you choose a strategy or predict a net result of several trials, you must balance the likelihood of each outcome against the payoff if it occurs. That is, a less likely outcome may be preferred to a more likely one if the payoff is correspondingly better.

key idea

We calculate the **mean** (symbolized by the Greek letter μ) for a probability model to determine the "average" payoff for the outcomes in the sample space.

question

If you repeatedly toss a coin and are offered $2 for each head and $1.50 for each tail, what would be your average expected winnings on each toss?

answer

$1.75

explanation

Here is the probability model:

Outcome of toss	H	T
Probability of outcome	.5	.5
Payoff for outcome	$2	$1.50

Thus the mean is calculated as follows:

$$\mu = 2 \times .5 + 1.5 \times .5 = 1.75$$

key idea

According to the **law of large numbers,** as the number of trials of a random phenomenon increases, the probability model describes more and more accurately the proportion of each outcome, and the mean μ approximates more closely the observed average values.

Sampling Distributions

[text pp. 265–268]

key idea

The opinions and characteristics of a given random sample of 100 people will differ somewhat from those of any other random sample. This is called **sampling variability.**

key idea

Any **statistic** that you calculate from samples, such as a mean value or a proportion of the sample with a given response or measurement, will vary with the samples in a way described by the **sampling distribution.**

question

Two pollsters are testing public opinion on a pending environmental bill. The first one samples groups of 50 people in various localities and records the percentage of each sample that supports the bill. The second does the same with samples of size 500. Describe and sketch a likely comparison between the histograms of their sampling distributions.

answer

Both distributions are symmetrically arranged around the mean—the true population percentage—which, for the purposes of illustration, is taken to be 40%. The first distribution, with smaller sample size, has greater spread around the mean. With larger samples, the distribution is more tightly centered. The distributions might look something like this:

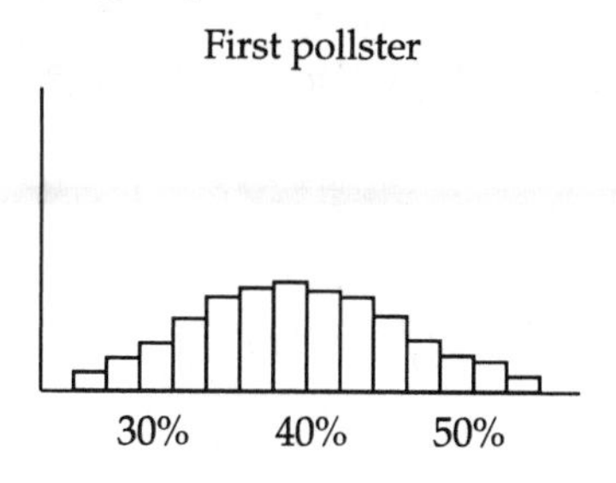

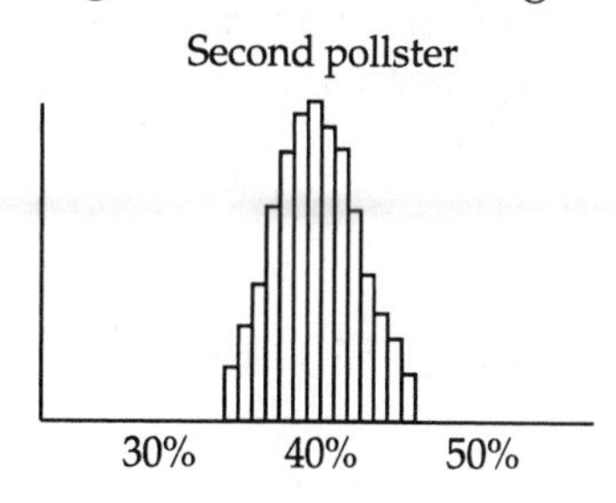

Normal Distributions

[text pp. 268–270]

key idea

Sampling distributions, and many other types of probability distributions, approximate a bell curve in shape and symmetry. This kind of shape is called a normal curve, and can represent a **normal probability distribution,** in which the area of a section of the curve over an interval coincides with the probability of an outcome within the interval.

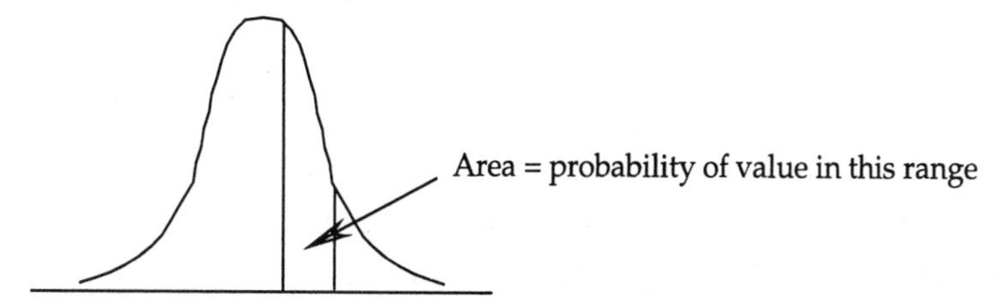

The Shape of Normal Curves

[text pp. 270–273]

key idea

The **mean** of a normal distribution is the center of the curve.

key idea

The **standard deviation** of a normal distribution is the distance from the mean to the point on the curve where the curvature changes.

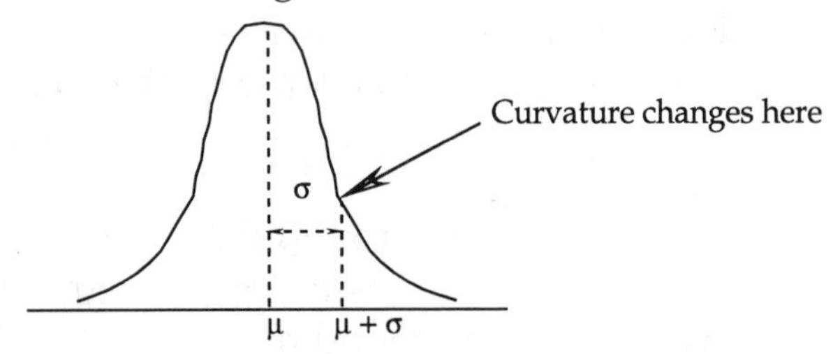

key idea

A normal distribution curve is completely determined by its mean and standard deviation. The first and third quartiles are located 0.67σ below and 0.67σ above the mean:

$$Q_1 = \mu - .67s \text{ and } Q_3 = \mu = .67\sigma$$

question

The scores on a Chemistry 101 test were normally distributed with a mean of 73 and a standard deviation of 12.

(a) Find the third quartile (Q_3) for the test scores.
(b) Find a range containing exactly half of the students' scores.

answer

(a) $Q_3 = 81$
(b) the interval [65, 81]

explanation

With $\mu = 73$ and $\sigma = 12$, $Q_3 = 73 + (.67 \times 12) = 81$ and $Q_1 = 73 - (.67 \times 12) = 65$.

Exactly half the scores lie between *Q1* and Q_3, which is the range [65, 81].

The 68–95–99.7 Rule

[text pp. 273–275]

key idea

The **68–95–99.7 rule** applies to a normal distribution. It is useful in determining the proportion of a population with values falling in certain ranges.

key idea

For a normal curve, these rules apply:

The proportion of the population within one standard deviation of the mean is 68%.
The proportion of the population within two standard deviations of the mean is 95%.
The proportion of the population within three standard deviations of the mean is 99.7%.

question

Look again at the Chemistry 101 test in which scores were normally distributed with a mean of 73 and a standard deviation of 12.

(a) Find a range containing 34% of the students' scores.
(b) What percentage of the test scores were between 61 and 97?

answer

(a) Either of the intervals [61, 73] or [73, 85]
(b) 81.5%

explanation

According to the given data and the 68–95–99.7 rule, the student scores fall into ranges as shown by this curve:

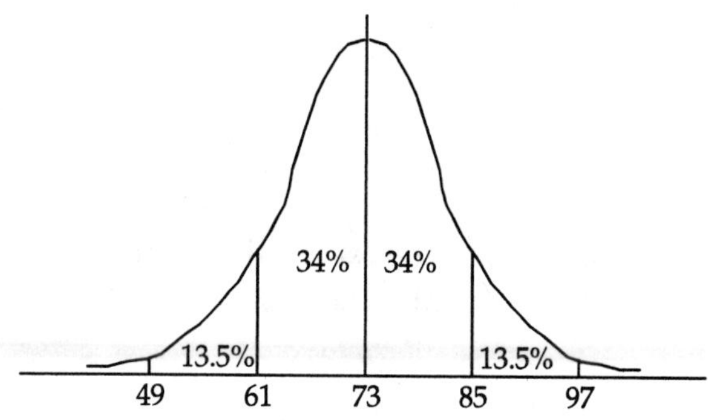

The Central Limit Theorem

[text pp. 275–277]

key idea

The reason that the normal distribution is so important is that for almost any random phenomenon, the mean value for random samples will tend to be normally distributed. This is the first principle contained in the **central limit theorem.**

key idea

The central limit theorem also says that a sample distribution will have the same mean μ as the original phenomenon, and will have a standard deviation equal to $\sigma / \sqrt{n}$, where σ is the standard deviation of a single trial and n is the number of trials.

question

Look one more time at the Chemistry 101 test in which scores were normally distributed with a mean of 73 and a standard deviation of 12.

(a) Suppose you chose 10 of the students at random and computed the mean $\bar{x}$ of their scores. What is the standard deviation of $\bar{x}$?
(b) How large a sample size n would you have to use to bring the standard deviation of $\bar{x}$ down to 2?

answer

(a) 3.8
(b) 36 or more

explanation

(a) Take the standard deviation of the original distribution, which is 12, and divide by the square root of 10; we get $12 / (3.16) = 3.8$.
(b) To bring it down from 12 to 2 or less, we would have to divide by 6 or more. If the square root of the sample size n is 6 or greater, then n must be at least 36.

Applying The Central Limit Theorem

[text pp. 277–281]

key idea

The central limit theorem can help to explain the advantage that a casino has over a gambler —over time, the gambler is almost certain to lose and the casino is guaranteed profit based on how many gamblers place bets. The more money that is bet, the more accurately a casino can predict its profits.

PRACTICE QUIZ

1. We will roll a die and flip two coins. Then we will report the number on the die and whether the coins are heads, tails, or mixed. How many outcomes are in the sample space?
   a. 8
   b. 9
   c. 18

2. Suppose we toss three coins and report the number of heads that appear. How many outcomes are in the sample space?
   a. 3
   b. 4
   c. 8

3. A sample space contains three outcomes: A, B, C. Which of the following could be a legitimate assignment of probabilities to the outcomes?
   a. P(A) = 0.4 P(B) = 0.6 P(C) = 0
   b. P(A) = 0.3 P(B) = 0.3 P(C) = 0.3
   c. P(A) = 0.6 P(B) = –0.2 P(C) = 0.6

4. We roll two dice and report the sum of the numbers rolled. The outcomes in this space are
   a. all equally likely.
   b. not all equally likely.

5. There are 5 red towels and 8 blue towels in a basket. We reach in and pull one out. What is the probability it is red?
   a. 5/8
   b. 5/13
   c. 1/2

6. A bicycle chain has a 4-digit code lock. How many possible codes are there?
   a. 40
   b. 5,040
   c. 10,000

7. Each raffle ticket costs $2. Of 400 tickets sold, one will win $250, and two others will each win $25. What is your mean value for one play?
   a. –$1.25
   b. –$0.75
   c. $0.75

8. The working life of a watch battery is 2 years, with a standard deviation of .75 years. What is the upper quartile of battery life?
   a. 2.5 years
   b. 2.75 years
   c. 3.5 years

9. The working life of a watch battery is 2 years, with a standard deviation of .75 years. What is the probability that it will last more than 6 months?
   a. 95%
   b. 97.5%
   c. more than 99%

10. The working life of a watch battery is 2 years, with a standard deviation of .75 years. Suppose a sample of 9 is drawn from a production run and tested. What is the standard deviation $\sigma_{\bar{x}}$ of the mean result?
    a. 0.75 years
    b. 0.083 years
    c. 0.289 years

CHAPTER 8

STATISTICAL INFERENCE

CHAPTER OBJECTIVES

Check off these skills once you feel you have mastered them.

- ❑ Define statistical inference.
- ❑ Describe the difference between informal and formal statistical inference.
- ❑ Explain the difference between a parameter and a statistic.
- ❑ Identify both the parameter and the statistic in a simple inferential setting.
- ❑ Compute the sample proportion when both the sample size and number of favorable responses are given.
- ❑ Using an appropriate formula, calculate the standard deviation of a given statistic.
- ❑ Compute the margin of error for a sample statistic when its standard deviation is given.
- ❑ Calculate differing margins of error for increasing sample sizes.
- ❑ Discuss the effect of an increased sample size on the statistic's margin of error.
- ❑ When given the standard deviation for a sample statistic, construct a 95% confidence interval for the population proportion.
- ❑ Explain the difference between the population mean and the sample mean.
- ❑ Discuss the use of process control charts in a manufacturing environment.
- ❑ Identify the center line and the control limits on a given process control chart.

GUIDED READING

Introduction

[text pp. 294–296]

Inference is the process of reaching conclusions from evidence. When we collect data to study a social or scientific issue, we display the data in various ways and calculate statistical measures of central tendency and spread. We then try to draw conclusions about characteristics of the population being studied. In this chapter, we shall use the principles of probability to determine how confident we are that our inferences are reasonably accurate. We rely on the mathematical properties of random sampling and normal distributions to extract confident conclusions from chance observations.

Estimating a Population Proportion

[text pp. 296–299]

key idea

We cannot know precisely a true population **parameter,** such as the proportion p of people who favor a particular political candidate. To make an estimate, we interview a random sample of the population and calculate a **statistic** of the sample, such as the **sample proportion** (written $\hat{p}$) favoring the candidate in question.

question

Suppose you conduct a telephone poll of 1250 people, asking them whether or not they favor mandatory sentencing for drug related crimes. If 580 people say "yes," what is the sample proportion $\hat{p}$ of people in favor of mandatory sentencing?

answer

$\hat{p} = 46.4\%$

explanation

$580 / 1250 = .464 = 46.4\%$

key idea

The sample proportion $\hat{p}$ (for a given sample size n) varies with the sample and falls into a normal sampling distribution, with mean = p, and **standard deviation** given by this **formula:**

$$\sigma_{\hat{p}} = \sqrt{\frac{p(100-p)}{n}}$$

question

Suppose that in the political poll from the previous question, the true population proportion is $p = 45\%$. What is the standard deviation of the sampling distribution?

answer

$\sigma_{\hat{p}} = 1.4$

explanation

Use the formula

$$\sigma_{\hat{p}} = \sqrt{\frac{p(100-p)}{n}}$$

Here $p(100 - p) / n = 45(55) / 1250 = 1.98$; take the square root to get 1.4.

key idea

The formula for $\sigma_{\hat{p}}$ shows that the spread of the sampling distribution is about the same for most sample proportions; it depends primarily on the sample size.

Confidence Intervals

[text pp. 299–304]

key idea

If a sample is large, the sample statistic is likely to be close to the true population parameter. If we take an interval $\pm 2\sigma_{\hat{p}}$ around the sample proportion $\hat{p}$, there is a 95% probability that the true proportion lies within that range. This is called a 95% confidence interval.

key idea

Since we do not know the value of p, we use $\hat{p}$ in the formula for $\sigma_{\hat{p}}$ and call the result $\hat{\sigma}_{\hat{p}}$.
Thus

$$\hat{\sigma}_{\hat{p}} = \sqrt{\frac{\hat{p}(100 - \hat{p})}{n}}$$

And a 95% confidence interval is given by $\hat{p} \pm 2\hat{\sigma}_{\hat{p}}$

key idea

If you are rounding off $\sigma_{\hat{p}}$, do not round down to make the confidence interval noticeably smaller.

question

In a political poll, 695 potential voters are asked if they have decided yet which candidate they will vote for in the next election. Suppose that 511 say "yes."
(a) Estimate the proportion of *undecided* voters.
(b) Find a 95% confidence interval for this estimate.

answer

(a) 26.5%
(b) 26.5% ± 3.4%; rounding off we get an interval (23%, 30%).

explanation

(a) The sample proportion of voters who have made up their minds is 511 / 695 = 73.5%, so the proportion of undecided voters is $\hat{p} = 26.5\%$.

(b)
$$\hat{p} \pm 2\,\hat{\sigma}_{\hat{p}} = 26.5 \pm 2\sqrt{\frac{26.5\,(73.5)}{695}} = 26.5 \pm 2\,(\sqrt{2.8}) = 26.5\% \pm 3.4\%.$$

key idea

To gain greater certainty in your statistical estimate, you must increase the confidence interval accordingly. That is, precision goes down as certainty goes up. To avoid this loss of precision, you must use a larger sample size in your poll.

Estimating a Population Mean

[text pp. 304–308]

key idea

We can use the **sample mean** (x) of a set of observations to estimate the mean of a

population (μ), with a confidence interval defining our level of certainty in the accuracy of our estimate.

key idea

If a population has a normal distribution with a mean μ and standard deviation σ, then the sampling distribution for samples of size n is approximately normal, with the same mean μ as the population parameter and with standard deviation $\sigma / \sqrt{n}$.

question

Each year, all the applicants to a university's biochemistry program take a mathematics placement test. Last year's scores formed a normal distribution with mean 18 and standard deviation 2.5. How would sample averages for random groups of five students be distributed?

answer

A normal distribution with mean 18 and standard deviation 1.12.

explanation

The mean of the sampling distribution is the same as the general population mean, in this case, 18. The standard deviation is $\sigma / \sqrt{n}$, in this case, $2.5 / \sqrt{5} = 2.5 / 2.236 = 1.12$.

key idea

If σ is known, we can use the sample mean $\bar{x}$ to estimate the population mean μ. A 95% confidence interval for this estimate is

$$\bar{x} \pm 2\frac{\sigma}{\sqrt{n}}$$

question

Referring to the placement test from the previous question, suppose this year's scores are known to fall into a normal distribution with standard deviation 2.7, but with the mean not yet known. Suppose you choose five students at random and note that their scores are: 14, 21, 17, 23, 19. Use this data set to estimate the overall mean μ, and establish a 95% confidence interval for your estimate.

answer

We estimate $\mu = 18.8$; the confidence interval is $18.8 \pm 2.4 = (16.4, 21.2)$.

explanation

Given the data (14, 21, 17, 23, 19), we get a sample mean $\bar{x} = 18.8$. Now we use the formula for the 95% confidence interval to get

$$\bar{x} \pm 2\frac{\sigma}{\sqrt{n}} = 18.8 \pm 2\frac{2.7}{\sqrt{5}} = 18.8 \pm 2\,(1.2) = 18.8 \pm 2.4.$$

Statistical Process Control

[text pp. 308–311]

key idea

An efficient method for quality control in a manufacturing process is to take repeated small random samples of the product and keep track of the average variation of the sampled units from an ideal standard or target value.

key idea

We can visualize overall quality by plotting sample averages over a period of time, with the target value *m* represented in the graph by a horizontal line. A **control chart** will also display two dashed lines called control limits at $\mu \pm 2\sigma / \sqrt{n}$, representing a range that should contain 95% of the data points if the manufacturing process is not flawed.

key idea

This is what a control chart might look like for a process with target value = 50 and standard deviation $\sigma = 2.7$, with samples of size 5. Note that the control limits are set at $50 \pm 2(1.2)$; that is, at 47.6 and 52.4.

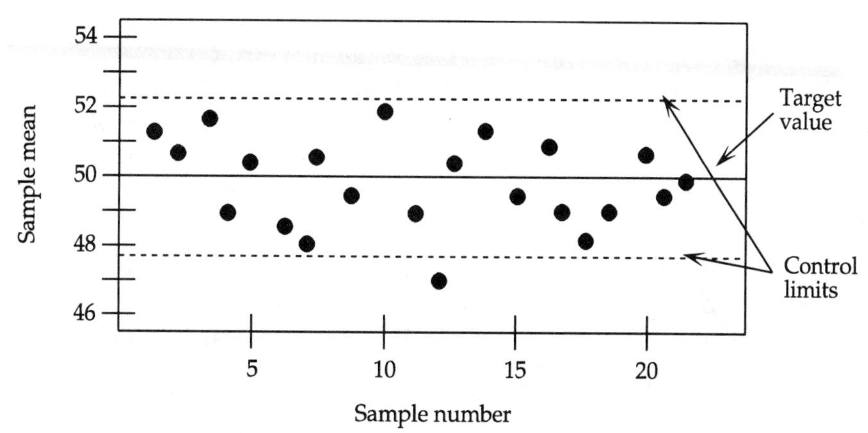

key idea

If the previous control chart were real, there would be no reason to be concerned about the quality of the process. Of the 22 samples taken, only one sample mean was outside the control limits. Thus, 95% of the samples are within appropriate standards of quality.

key idea

With industrial processes requiring a higher level of reliability, it is common to place the control limits three standard deviations away from the target value. This is more rigorous because, in this case, 99.7% of the samples would have to be within the standards to indicate a satisfactory process.

The Perils of Data Analysis

[text pp. 311–313]

key idea

Statistical evidence not based on experiments can show that an effect is present without showing *why* the effect is present. Sometimes, looking at data in a **two-way table** can help to identify the reason for the effect.

Even when data is produced and analyzed carefully, there is always the chance that the conclusions drawn from the data are incorrect. The strength of statistical inference is in setting confidence levels as high as we deem necessary.

PRACTICE QUIZ

1. A random sample of 10 bags of flour has a mean weight of 24.9 pounds, less than the mean weight 25.05 pounds of all bags produced. In this example. 25.05 is called a
   a. statistic.
   b. parameter.
   c. sample.

2. A random sample of 10 bags of flour has a mean weight of 24.9 pounds, less than the mean weight 25.05 pounds of all bags produced. In this example, 24.9 is called a
   a. statistic.
   b. parameter.
   c. sample.

3. To determine interest in a new paper towel, samples are received and tried by 300 local residents, of which 120 prefer it to their current brand. What is the sample proportion $\hat{p}$?
   a. 12%
   b. 40%
   c. 2.5 %

4. To determine interest in a new paper towel, samples are mailed to 300 local residents, of which 120 prefer it to their current brand. What is the standard deviation of the sampling distribution of this statistic?
   a. 2.83%
   b. 3.65%
   c. 4.47%

5. Bags of flour produced by a company have a mean weight of 25.5 pounds and a standard deviation of .5 pounds. Four bags are chosen at random and a mean weight is found. What is the standard deviation of the sampling distribution for the mean weight?
   a. 0.125 pound
   b. 0.25 pound
   c. 0.5 pound

6. A random poll of 600 people shows that 60% of those polled are in favor of a new school building. Find a 95% confidence interval for the proportion of the residents in favor of a new school building.
   a. 56% to 64%
   b. 40% to 80%
   c. 58% to 62%

7. A random sample of 350 freshmen at a local high school found that 25 have never been to a dentist. Find a 95% confidence interval for the true proportion of freshmen at this school that have never been to a dentist.
   a. 7.1% ±1.4%
   b. 7.1% ±1.9%
   c. 7.1% ±2.8%

8. A control chart is used to
   a. display the five-number summary for a set of quality control data.
   b. present a probability model.
   c. show if a product process is operating within acceptable bounds.

9. Bags of flour produced by a company have a mean weight of 25.5 pounds and a standard deviation of .5 pounds. The company uses 95% control limits for a process control chart for this procedure. A random sample of 16 bags is selected and weighed, with a mean sample weight of 25.8 pounds. Which of the following statements is true?
   a. The sample mean weight is out of control.
   b. The sample mean weight is not out of control.
   c. There is not enough information given.

10. Here is a two-way table for admission into a private club. Find the percentage of applicants admitted.

	Number of males	Number of females
Admitted	20	20
Denied	30	10

   a. 50%
   b. more than 50%
   c. less than 50%

CHAPTER 9

IDENTIFICATION NUMBERS

CHAPTER OBJECTIVES

Check off these skills once you feel you have mastered them.

- ❑ Understand the purpose of a check digit and be able to determine one for various schemes.
- ❑ Given an identification number and the scheme used to determine it, be able to decide if the number is a valid number for that scheme.
- ❑ Be able to convert a given ZIP code to its corresponding bar code, and vice versa.
- ❑ Be able to convert a given UPC number to its corresponding bar code.

GUIDED READING

Introduction

[text pp. 331–339]

Almost everything we encounter in daily life—consumer goods, credit cards, financial records, people, organizations, mail—is somehow identified or classified by a numeric or alphanumeric code. Of course, this code must unambiguously identify the individual or object it names. But since humans and machines are fallible, the system used for creating the code must be designed to minimize errors. Also, since errors will certainly occur, the system should include a mechanism for detecting and, if possible, correcting the most common errors.

key idea

Many frequently used types of **error-detecting code** for identification numbers include an extra digit (usually the last digit) called a check digit. The check digit can be compared to the rest of the number to check for validity.

key idea

For a postal service money order, the check digit (the last digit) is the remainder you get when you divide the sum of the other digits by 9.

question

If the ID number on a postal money order is 321556738X, what is the value of X?

answer

X = 4

explanation

The other digits add up to 40; divide by 9, the remainder is 4.

key idea

Some mail and car rental services use as an extra check digit the remainder when you divide the number by 7.

question

If the ID number of a FedEx package is 321556738X, what is the value of X?

answer

X = 6

explanation

If you divide 321556738 by 7, you get 45936676 with a remainder of 6.

key idea

The *Universal Product Code (UPC)* adds a check digit at the end by adding the digits and multiplying by their **weight** (alternately 1 for even positions, 3 for odd positions). The sum must be a number ending in 0.

question

If the UPC code for a product is 432765987X, what is the value of X?

answer

X = 3. The correct code is 4327659873.

explanation

The calculation works like this:

$$(4+2+6+9+7)\bullet 3+(3+7+5+8+X)\bullet 1=84+23+X=107+X,$$

so X = 3.

107 + 3 = 110, a number ending in 0.

key idea

The **Codabar** system is a variation of UPC using a similar sum with **weights** 2 (odd positions) and 1 (even positions). To this sum, you add the number of digits in odd positions that exceed 4; the resulting number must end in 0 to be a valid Codabar code.

question

If the Codabar code for a credit card is 432765987X, what is the value of X?

answer

X = 8 and the correct code is 4327659878.

explanation

The calculation works like this:

$$(4+2+6+9+7)\bullet 2+(3+7+5+8+X)\bullet 1+3=56+23+X+3=82+X,$$

so X = 8.

82 + 8 = 90, a number ending in 0.

key idea

Some other important and effective error-detecting codes are the **International Standard Book Number (ISBN)** and **Code 39**.

The ZIP Code

[text pp. 339–341]

key idea

The **ZIP code** is a U. S. Postal Service ID that **encodes** geographical information about each post office.

question

What does the first digit in a ZIP code mean?

answer

The country is divided into ten regions, from east to west, numbered 0–9. The first digit represents one of these areas.

question

What do the second and third digits in a ZIP code mean?

answer

Each state is divided into a variable number of smaller geographical areas. The second two digits represent the central mail-distribution point in this area.

key idea

ZIP + 4 code is a further refinement of the ZIP code, capable of identifying small groups of mailboxes, like a floor of a building, within a given postal zone.

Bar Codes

[text pp. 341–350]

key idea

Bar codes use light spaces and dark bars to represent a two-symbol **binary code** that is easily scanned optically and **decoded** by a computer.

key idea

ZIP code bar codes use the **postnet code**. Each digit is represented by a group of five dark bars, two long and three short. There are nine digits for the ZIP + 4 number and one check digit.

question

Suppose the first three digits of a ZIP code are 194 . . . What would the bar code look like?

answer

explanation

In groups of 5;

ııı|| = 1, |ı|ıı = 9, ı|ıı| = 4

key idea

The **UPC** (Universal Product Code) is a familiar sight on labels for retail products. Digits are represented by sequences of light and dark bars, where adjacent dark bars blend together to make bars of different widths. In this way, seven bar spaces ("modules") produce two light and two dark bars for each digit. There are different binary coding patterns for manufacturer numbers and product numbers.

question

How is the digit 7 represented in the UPC bar code?

answer

For manufacturer code, 7 = 0111011.

For product code, 7 = 1000100.

The bar patterns would look something like this:

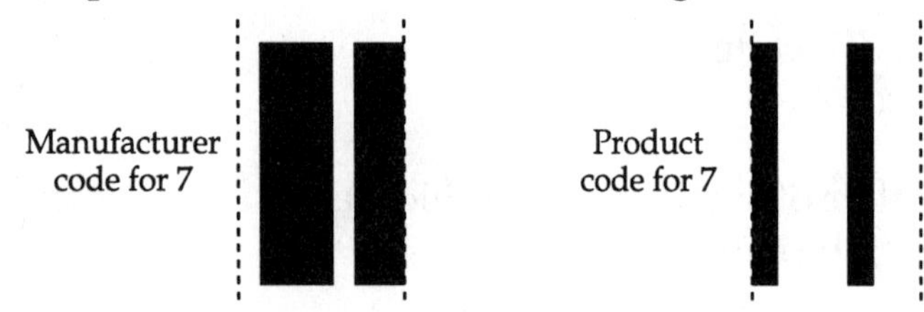

PRACTICE QUIZ

1. Suppose a U. S. Postal Service money order is numbered 632930421#, where the last digit is obliterated. What is the missing digit?
   a. 3
   b. 6
   c. 0

2. Suppose an American Express Travelers Cheque is numbered #293019225, where the first digit is obliterated. What is the missing digit?
   a. 3
   b. 4
   c. 6

3. Is the number 3281924 a legitimate Avis rental car number?
   a. Yes
   b. No, but if the final digit is changed to a 2, the resulting number 3281922 is legitimate.
   c. No, but if the final digit is changed to a 3, the resulting number 3281923 is legitimate.

4. Is the number 193234580012 a legitimate airline ticket number?
   a. Yes
   b. No, but if the final digit is changed to a 4, the resulting number 193234580014 is legitimate.
   c. No, but if the finial digit is changed to a 1, the resulting number 193234580011 is legitimate.

5. Determine the check digit that should be appended to the UPC code 0-10010-34500.
   a. 2
   b. 4
   c. 8

6. Determine the check digit that should be appended to the bank identification number 015 000 64.
   a. 2
   b. 5
   c. 8

7. Determine the check digit that should be appended to the Codabar number 312580016535003.
   a. 1
   b. 3
   c. 7

8. Suppose the ISBN 0-1750-3549-0 is incorrectly reported as 0-1750-3540-0. Which of the following statements is true?
   a. The check digit will detect the error, but cannot correct it.
   b. The check digit will detect and correct the error.
   c. The check digit cannot detect the error.

9. Determine the check character that should be appended to the Code 39 number 10000ACD349110.
   a. 8
   b. S
   c. another character

10. Suppose that a Postnet code is incorrectly reported. You know that only one of the digits is incorrectly reported. Which of these statements is true?
    a. If you know which digit is incorrect, you can always correct a single error in a Postnet code.
    b. If you know which digit is incorrect, you can sometimes but not always correct a single error in a Postnet code.
    c. Even if you know which digit is incorrect, you can never correct a single error in a Postnet code.

CHAPTER 10

TRANSMITTING INFORMATION

CHAPTER OBJECTIVES

Check off these skills once you feel you have mastered them.

- ❑ Know what a binary code is.
- ❑ Be able to compute check digits for code words given the parity check sums for the code.
- ❑ Be able to determine the distance between two n-tuples of 0's and 1's.
- ❑ Be able to determine the weight of a code word and the minimum weight of the nonzero code words in a code (for binary codes, the minimum weight is the same as the minimum distance between code words).
- ❑ Know what nearest-neighbor decoding is and be able to use it for decoding messages received in the Hamming code of Table 10.2.

GUIDED READING

Introduction

[text p. 358]

In this chapter we consider some sophisticated techniques to detect and correct errors in digitally transmitted messages. We also take a look at methods that have been developed for the compression of data and for protection of the confidentiality of our messages.

Binary Codes

[text pp. 358–361]

key idea

Most computerized data are stored and transmitted as sequences of 0's and 1's. If we imagine storing each digit in a compartment, then we can detect and even correct errors by using several check digits to record the **even or odd parity** of the compartments' total data.

question

Store the data string 1011 along with its check digits correctly in a three-circle diagram, using the scheme from the textbook, as shown here:

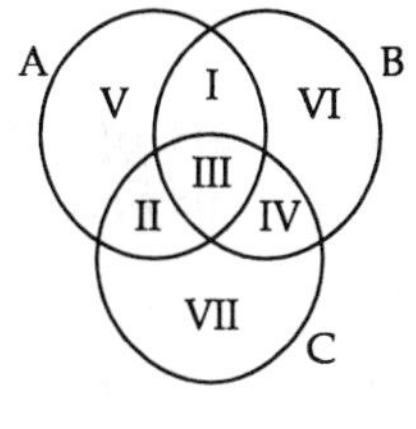

A B C I II III IV V VI VII

answer

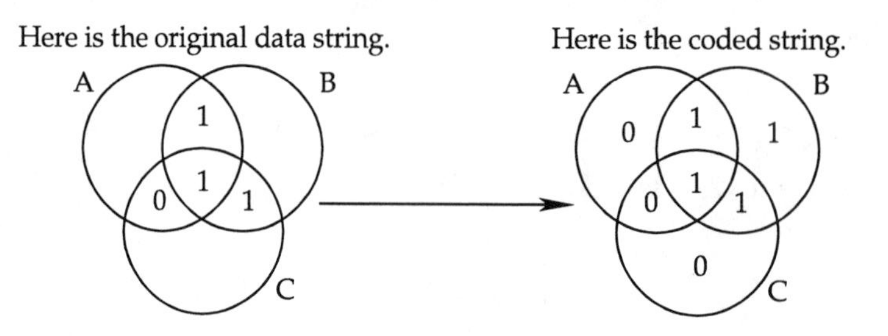

explanation

Note that the parity of each circle A, B, C is even.

Encoding with Parity-Check Sums

[text pp. 361–370]

key idea

If a string of digits $a_1\,a_2\,a_3\,a_4$ is a message, then we can create a **code word** in a **binary linear code** by adding check digits that are the **parity-check sums** $a_1 + a_2 + a_3$, $a_1 + a_3 + a_4$, and $a_2 + a_3 + a_4$.

question

What is the code word for the message 1100?

answer

1100011

explanation

$a_1 + a_2 + a_3 = 2$, even parity; so the first check digit is 0. Both $a_1 + a_3 + a_4$ and $a_2 + a_3 + a_4 = 1$, odd parity; so the last two check digits are 1.

key idea

Since there may be errors in the coding process or transmission, we use the **nearest-neighbor decoding** method to find the code word with the shortest **distance** (number of positions in which the strings differ) from the received message.

question

Using the nearest-neighbor method, decode these two code words.

a) 1110110
b) 0110010

answer

a) 1110110 → 1110100
b) 0110010 → 0110010

explanation

a) The distance between the strings is 1, the error is in the sixth digit.
b) The distance is 0, there is no error. The message is a valid code word.

key idea

The **weight of a code** determines how many check digits are needed for error correction.

key idea

In a **variable-length** code, we can use **data compression** to make the shortest code words correspond to the most frequently occurring strings. Morse code is an example.

Cryptography

[text pp. 370–376]

key idea

Encryption of stored and transmitted data protects its security. For example, **passwords** are stored in encrypted form in computers.

key idea

Cable television companies verify your key as a valid customer before unscrambling the signal. They use a method of matching strings by "addition": 0 + 0 = 0, 0 + 1 = 1, 1 + 1 = 0. Two strings match if they "add" to a string of 0's.

question

If your **key** is $k = 10011101$ and the transmitted message is $p + k = 01101100$, what is the password p that will unscramble your signal?

answer

11110001

explanation

Add the strings 10011101 + 01101100 to get 11110001.

$$\begin{array}{r} 10011101 \\ \underline{01101100} \\ 11110001 \end{array}$$

Place a 0 wherever the strings match and a 1 wherever they do not match.

PRACTICE QUIZ

1. If you use the circular diagram method to encode the message 1010, what is the encoded message?
   a. 1010001
   b. 1010010
   c. 1101000

2. Suppose the message 1111100 is received and decoded using the nearest-neighbor method. What message is recovered?
   a. 1111
   b. 1110
   c. 0111

3. What is the distance between received words 1111100 and 1101101?
   a. 5
   b. 4
   c. 2

4. For the code C = {0000, 1010, 0101, 1111}, how many errors would have to occur during the transmission for a received word to possibly be encoded incorrectly?
   a. 1
   b. 2
   c. 3

5. Use the encoding scheme A → 0, B → 10, C → 11 to decode the sequence 0110011.
   a. ABBAC
   b. ACBAC
   c. ACAAC

6. What is the sum of the binary sequences 1001101 and 1100011?
   a. 1101111
   b. 0101110
   c. 0110000

7. Using modular arithmetic, 3^6 mod 20 is equal to
   a. 3.
   b. 9.
   c. 19.

8. For the RSA scheme with $p = 11$ and $q = 17$, which of the following could be chosen as a value for r?
   a. 5
   b. 6
   c. 9

9. For the RSA scheme with $m = 5$ and $r = 8$, what is the value of s?
   a. 2
   b. 3
   c. 5

10. Use the RSA scheme with $n = 91$ and $s = 5$ to decode the message "6".
    a. 41
    b. 64
    c. 85

CHAPTER 11

SOCIAL CHOICE: THE IMPOSSIBLE DREAM

CHAPTER OBJECTIVES

Check off these skills when you feel that you have mastered them.

- ❑ Analyze and interpret a preference schedule.
- ❑ Rearrange a preference schedule to accommodate the elimination of one or more alternatives.
- ❑ Explain the difference between majority rule and the plurality method.
- ❑ Discuss why the majority method may not be appropriate for an election in which there are more than two candidates.
- ❑ Apply the plurality voting method to determine the winner in an election whose preference schedule is given.
- ❑ Determine the Borda count winner from a preference schedule.
- ❑ Structure two alternative contests from a preference schedule by rearranging the schedule; then determine whether a Condorcet winner exists.
- ❑ Describe the process of insincere voting.
- ❑ Discuss Condorcet winner criterion (CWC).
- ❑ Discuss independence of irrelevant alternatives (IIA).
- ❑ Discuss Pareto condition.
- ❑ Discuss monotonicity.
- ❑ Discuss Arrow's impossibility theorem.
- ❑ List three factors that can affect the outcome of an election.
- ❑ Recognize the application of the law of transitivity in the interpretation of individual preference schedules and its possible nonvalidity in group preferences (the Condorcet paradox).
- ❑ Explain the process of approval voting and discuss its consideration in political races.

GUIDED READING

Introduction

[pp. 385–387]

Voting occurs in many situations, such as in elections of public officials, officers of a club, or among a group of friends who have to decide in which restaurant to eat. While elections involving just two choices are quite simple, the opposite is true for elections with three or more alternatives, in which many complications and paradoxes arise. Social choice theory was developed to analyze the various types of voting methods, to discover the potential pitfalls in each, and to attempt to find improved systems of voting.

Elections with Only Two Alternatives

[text p. 387]

key idea

When there are only two candidates or alternatives, **May's theorem** states that **majority rule** is the only voting method that satisfies three desirable properties, given an odd number of voters and no ties.

question

State the three properties satisfied by majority rule.

answer

The three properties are:

1. All voters are treated equally.
2. Both candidates are treated equally.
3. If a single voter who voted for the loser, *B*, changes his mind and votes for the winner, *A*, then *A* is still the winner.

Elections with Three or More Alternatives

[text pp. 388–398]

key idea

In plurality voting, the candidate with the most first-place votes on the preference lists is the winner. We do not take into account the voters' preferences for the second, third, etc., places.

question

Find the winner for the following preference list under **plurality voting**. (The first row of the table gives the number of voters for each particular preference list.)

31	22	20	27
A	*B*	*B*	*C*
B	*C*	*A*	*A*
C	*A*	*C*	*B*

answer

The answer is *B*.

explanation

Adding the first-choice votes in the second and third columns, we see that *B* has a total of 42 votes, which is a majority.

key idea

A candidate is the **Condorcet winner** if she defeats every other candidate in head-to-head elections.

question

In the following table, is there a Condorcet winner? If so, who is it?

31	22	20	27
A	*B*	*B*	*C*
B	*C*	*A*	*A*
C	*A*	*C*	*B*

answer

A is the Condorcet winner.

explanation

A beats *B* 58 to 42, and *A* beats *C* 51 to 49.

question

In plurality voting, must there always be a Condorcet winner?

answer

The answer is no.

explanation

In the previous example, reduce the number of voters in the first column from 31 to 29, and increase the number in the second column to 24. We thus obtain a new preference list:

29	24	20	27
A	*B*	*B*	*C*
B	*C*	*A*	*A*
C	*A*	*C*	*B*

Then *A* beats *B* 56 to 44, *B* beats *C* 73 to 27, and *C* beats *A* 51 to 49, so that there is no Condorcet winner.

key idea

In the **Borda count** method, points are assigned to each position in the set of preference lists. For example, in a 3-person election, first-place votes may be awarded 2 points each, second-place votes receive 1 point each, and third-place votes are given 0 points each. (Other distributions of points may be used to create similar rank methods.)

question

Determine the Borda count winner in the following table, using a 2-1-0 point distribution.

31	22	20	27
A	*B*	*B*	*C*
B	*C*	*A*	*A*
C	*A*	*C*	*B*

answer

B is the winner.

explanation

A has 31 first-place votes, 47 second-place votes, and 22 third-place votes. He thus has a total of $(31)(2) + (47)(1) + (22)(0) = 109$ points. Similarly, B has $(42)(2) + (31)(1) + (27)(0) = 115$ points, while C has $(27)(2) + (22)(1) + (51)(0) = 76$ points.

key idea

The Borda count method may be subject to **strategic voting**.

question

If all of the voters vote sincerely, who is the winner in the following preference list, using a 3-2-1 point distribution?

7	3	3
A	B	C
B	A	B
C	C	A

answer

A is the winner.

explanation

A has $(7 \times 3) + (3 \times 2) + (3 \times 1) = 30$ points, B has $(3 \times 3) + (10 \times 2) = 29$ points, while C has $(3 \times 3) + (10 \times 1) = 19$ points.

question

Suppose the three voters in the last column (those whose preference is CBA) vote strategically, by switching their preferences to BCA. Rewrite the preference lists and determine who will now be the winner.

answer

B is the winner.

explanation

The new preference list is:

7	3	3
A	B	B
B	A	C
C	C	A

A still has 30 points, but B now has $(6 \times 3) + (7 \times 2) = 32$ points, and C has only $(3 \times 2) + (10 \times 1) = 16$ points.

key idea

An **agenda** is the listing (in some order) of the alternatives. **Sequential pairwise voting** pits the first alternative against the second in a head-to-head contest. The winner goes on to confront the third alternative, while the loser is eliminated. The alternative remaining at the end is the winner.

question

Who is the winner in the following table, if the order of the agenda is *A*, *B*, *C*, and *D*; that is, if *A* competes against *B* first, then the winner against *C*, and so forth?

5	2	4
A	*B*	*C*
B	*C*	*D*
D	*A*	*A*
C	*D*	*B*

answer

C is the winner.

explanation

A beats *B* 9 to 2, then loses to *C* 6 to 5. Finally, *C* beats *D* 6 to 5, so that *C* is the winner.

key idea

The choice of the agenda can affect the result.

question

The chart from the previous investigation is shown below. Who wins with the agenda *BCDA*?

5	2	4
A	*B*	*C*
B	*C*	*D*
D	*A*	*A*
C	*D*	*B*

answer

A is the winner.

explanation

B beats *C* 7 to 4, then beats *D* by the same margin. Finally, *A* beats *B* 9 to 2.

key idea

Sequential pairwise voting fails to satisfy the **Pareto condition**, which states that if everyone prefers one alternative, say *A*, to another, say *B*, then *B* cannot be the winner.

key idea

In the **Hare system**, the winner is determined by repeatedly deleting alternatives that are the least preferred, in the sense of being at the top of the fewest preference lists.

question

Apply the Hare system to the table shown below and determine the winner.

5	2	4
A	*B*	*C*
B	*C*	*D*
D	*A*	*A*
C	*D*	*B*

answer

C is the winner.

explanation

D has no first-place votes and is thus eliminated first. *B* has just 2 first-place votes and is eliminated next. At this point, with *B* and *D* eliminated, *A* heads 5 ballots, and *C* 6, so *C* is the winner.

key idea

The Hare system does not satisfy **monotonicity**.

Insurmountable Difficulties: From Paradox to Impossibility

[text pp. 398–405]

key idea

Condorcet's voting paradox occurs if, for example, in a three-candidate race, the voters favor *A* over *B*, *B* over *C*, and *C* over *A*.

question

Does Condorcet's voting paradox occur in the following table?

5	2	4	2
A	*B*	*C*	*D*
B	*C*	*D*	*B*
D	*A*	*A*	*A*
C	*D*	*B*	*C*

answer

The answer is no.

explanation

A is the Condorcet winner. Therefore, there is no Condorcet paradox.

question

Does Condorcet's voting paradox occur in the following table?

29	24	20	27
A	*B*	*B*	*C*
B	*C*	*A*	*A*
C	*A*	*C*	*B*

answer

The answer is yes.

explanation

Voters prefer *A* over *B* (56 to 44) and *B* over *C* (73 to 27), but *C* over *A* (51 to 49).

key idea

Arrow's impossibility theorem states that there is no voting procedure that satisfies both the Condorcet winner criterion (CWC) and independence of irrelevant alternatives (IIA).

A Better Approach? Approval Voting

[text pp. 405–406]

key idea

In approval voting, each voter may vote for as many candidates as s/he chooses.

question

Who is the winner in the following table, where *X* indicates that the voter approves of that particular candidate?

Candidates	4	3	5	2
A	X		X	X
B		X	X	
C	X	X		
D				X
E	X	X	X	
F		X	X	

answer

E is the winner.

explanation

The vote totals were as follows:

A has 4 + 5 + 2 = 11 votes.
B has 3 + 5 = 8 votes.
C has 4 + 3 = 7 votes.
D has 2 votes.
E has 4 + 3 + 5 = 12 votes.
F has 3 + 5 = 8 votes.

PRACTICE QUIZ

1. Majority rule is an effective way to make a choice between
    a. two alternatives.
    b. a small number of alternatives.
    c. any number of alternatives.

2. The first-place votes for each of four candidates are counted, and the candidate with the most votes wins. This voting system is an example of
   a. majority rule.
   b. approval voting.
   c. plurality voting.

3. Each voter ranks the four candidates. The candidate who is ranked above any of the other candidates by a majority of the voters is declared to be the winner. This is an example of
   a. Condorcet winner criterion.
   b. Borda count.
   c. Hare system.

4. 11 committee members need to elect a chair from the candidates A, B, C, and D. The preferences of the committee members are given below. Which candidate will be selected if they use majority rule?

	6 members	2 members	3 members
1st choice	A	B	C
2nd choice	B	C	D
3rd choice	C	D	B
4th choice	D	A	A

   a. A
   b. B
   c. C

5. 11 committee members need to elect a chair from the candidates A, B, C, and D. The preferences of the committee members are given below. Which candidate will be selected if they use a Borda count?

	6 members	2 members	3 members
1st choice	A	B	C
2nd choice	B	C	D
3rd choice	C	D	B
4th choice	D	A	A

   a. A
   b. B
   c. C

6. 11 committee members need to elect a chair from the candidates A, B, C, and D. The preferences of the committee members are given below. Which candidate will be selected if they use a rank system which assigns 5 points to a first choice vote, 3 points for a 2nd place vote, 2 points for a 3rd place vote, and 1 point for a 4th place vote?

	6 members	2 members	3 members
1st choice	A	B	C
2nd choice	B	C	D
3rd choice	C	D	B
4th choice	D	A	A

   a. A
   b. B
   c. C

7. 37 members must elect a club president. Preferences among candidates A, B, C, and D are given below. Which candidate wins under the Hare system?

	14 members	10 members	8 members	4 members	1 member
1st choice	A	C	D	B	A
2nd choice	B	B	C	D	D
3rd choice	C	D	B	C	B
4th choice	D	A	A	A	C

   a. A
   b. C
   c. D

8. 37 members must elect a president of their club. Preferences among candidates A, B, and C are given below. Which candidate is the Condorcet winner?

	14 members	11 members	12 members
1st choice	A	B	C
2nd choice	B	C	B
3rd choice	C	A	A

   a. A
   b. B
   c. C

9. 25 partygoers have enough money together for a one-topping super-size party pizza. They each mark what toppings they find acceptable, as shown below. Which topping will be selected using approval voting?

	8 voters	6 voters	4 voters	4 voters	3 voters
pepperoni		X	X	X	
mushrooms				X	X
anchovies	X		X		X

   a. pepperoni
   b. mushrooms
   c. anchovies

10. Which voting method satisfies both the Condorcet winner criterion and the independence of irrelevant alternatives?
    I. Condorcet method
    II. Plurality
    a. Only I
    b. Only II
    c. Neither I nor II

CHAPTER 12

WEIGHTED VOTING SYSTEMS

CHAPTER OBJECTIVES

Check off these skills once you feel that you have mastered them

- ❑ Interpret the symbolic notation for a weighted voting system by identifying the quota and the number of votes each voter controls.
- ❑ Identify winning and losing coalitions by analyzing a given weighted voting system.
- ❑ Determine from the given notation whether a weighted voting system has a dictator.
- ❑ Calculate the number of coalitions for a given weighted voting system.
- ❑ When given a specific coalition from a weighted voting system, determine the critical voters in the coalition.
- ❑ List the eight coalitions for a three-voter weighted voting system.
- ❑ Calculate the Banzhaf power index for a given weighted voting system.
- ❑ Explain the difference between a winning coalition and a minimal winning coalition.
- ❑ List the possible permutations for a three- or four-voter weighted voting system.
- ❑ Calculate the Shapley-Shubik index for a three- or four-voter weighted voting system.
- ❑ Determine the extra votes for a winning coalition, and identify the critical voters.

GUIDED READING

Introduction

[text p. 416]

There are many settings, such as shareholder elections, in which people who are entitled to vote have varying numbers of votes. In such situations, the actual number of votes each can cast may not reflect the voter's *power*, that is, his ability to influence the outcome of the election. Several measures of power have been introduced and two of them are studied in this chapter, the *Banzhaf Power Index* and the *Shapley–Shubik Power Index*.

How Weighted Voting Works

[text pp. 416–421]

key idea

A **weighted voting system** is one in which each voter has a number of votes, called his or her **weight**. The number of votes needed to pass a measure is called the **quota**. If the quota for a system with n voters is q, and the weights are $w_1, w_2, \ldots w_n$, then we use this shorthand notation for the system: $[q: w_1, w_2, \ldots w_n]$.

key idea

A **winning coalition** is a combination of voters with enough collective weight to pass a measure.

question

List all of the winning coalitions in the weighted voting system given by $[q : w(A), w(B), w(c)] = [7 : 5, 3, 2]$.

answer

The winning coalitions are $\{A, B\}$, $\{A, C\}$, and all three voters, $\{A, B, C\}$.

explanation

The coalition $\{A, B\}$, with $5 + 3 = 8$ votes, exceeds the quota of 7.

The coalition $\{A, C\}$, with $5 + 2 = 7$ votes, matches the quota of 7.

The coalition $\{B, C\}$, with $3 + 2 = 5$ votes, is less than the quota of 7 and is thus a losing coalition.

key idea

A **dummy** is one whose vote will never be needed to pass or defeat any measure.

question

List any dummy in the weighted voting system $[q : w(A), w(B), w(C), w(D)] = [12 : 8, 6, 4, 1]$.

answer

The answer is D.

explanation

The winning coalitions that contain D are $\{A, B, D\}$, $\{A, C, D\}$, and $\{A, B, C, D\}$. In each case, if D were to drop out, the remaining members would still form a winning coalition.

key idea

A **dictator** is a voter whose weight is greater than or equal to the quota. A **blocking coalition** is a group of voters who have a sufficient number of votes to block a measure from passing. A one-person blocking coalition is said to have **veto power**.

question

List the blocking coalitions in the weighted voting system $[q: w(A), w(B), w(C), w(D)] = [12: 8, 6, 4,1]$. Does any voter have veto power?

answer

$\{A\}$, $\{A, D\}$, $\{B, C\}$ and $\{B, C, D\}$ are blocking coalitions. A has veto power.

explanation

$w = 8 + 6 + 4 + 1 = 19$ and $q = 12$. Then $w - q = 19 - 12 = 7$. Any losing coalition with more than 7 votes is a blocking coalition. Voter A has veto power because he alone has 8 votes, which is enough to block any measure.

The Banzhaf Power Index

[text pp. 421–431]

key idea

A voter is **critical** to a winning or blocking coalition if he can cause that coalition to lose by single-handedly changing his vote. Note that some winning coalitions have several critical voters, while others have none at all.

question

In the weighted voting system $[q : w(A), w(B), w(C), w(D)] = [12 : 8, 6, 4, 1]$, $\{A, B, D\}$ is a winning coalition. Find the critical voters in this coalition.

answer

A and B are critical voters.

explanation

If either A or B drops out, the votes of the remaining members of the coalition total less than 12. On the other hand, D is not critical, since A and B together have 14 votes.

key idea

A winning coalition with total weight w has $w - q$ **extra votes**. The critical voters in the coalition are those with weight greater than the extra votes number.

question

For the voting system [8: 5, 5, 3, 2] with voters A, B, C and D, the coalition $\{B, C, D\}$ is a winning coalition. How many extra votes does it have, and which are the critical voters?

answer

2 extra votes. B and C, with weights 5 and 3, are critical.

explanation

The coalition has votes $\{5, 3, 2\}$, for a total weight $w = 10$. Since the quota is $q = 8$, there are $w - q = 10 - 8 = 2$ extra votes. Since B (weight = 5) and C (weight = 3) have more votes than the extra votes number, they are critical. But D has only two votes and therefore is not critical.

key idea

A voter's **Banzhaf power index** equals the number of distinct winning coalitions in which he is a critical voter.

question

Find the Banzhaf power indices for the voters in the weighted voting system $[q : w(A), w(B), w(C), w(D)] = [12 : 8, 6, 4, 1]$.

answer

The answer is (12, 4, 4, 0).

explanation

The winning coalitions are

{A, B} {A, C} {A, B, C}
{A, B, D} {A, C, D} {A, B, C, D}

with the critical voters circled in each. Thus, for example, *A* is critical in 6 winning coalitions. He is also critical in 6 blocking coalitions (the numbers are always equal) so his total Banzhaf index is 12. Similar calculations may be made for other voters. Of course, *D*, who is a dummy, has an index of 0.

key idea

When there are many voters, the number of winning coalitions can be very large, and calculating the Banzhaf index will then be cumbersome. However, there are settings in which most, but not all, of the voters have equal weights. In such situations, we can compute the Banzhaf index by means of **combinations**, using the numbers C_k^n

question

Find C_3^7

answer

$$C_3^7 = \frac{7 \times 6 \times 5}{3 \times 2 \times 1} = 35.$$

question

Find the Banzhaf power indices for the voters in the weighted voting system $[q : w(A), w(B), w(C), w(D), w(E)] = [4 : 2, 1, 1, 1, 1]$.

answer

The answer is (20, 8, 8, 8, 8).

explanation

A is a critical voter in just two types of winning coalitions: *Axx* and *Axxx*, where *x* indicates any of the other voters. In the first situation, the two additional voters are drawn from the four other voters, and there are $C_2^4 = 6$ ways of choosing these two voters. Similarly, in the second case, there are $C_3^4 = 4$ ways of choosing three voters from among four. Hence, *A* is critical to 10 winning coalitions, and there are 10 blocking coalitions in which he is also critical, so that his Banzhaf index is 20. Now let us consider one of the voters with just one vote, say *B*, who is also critical in two types of winning coalitions: *ABx*, which can happen in three ways, since *x* can be *C*, *D* or *E*, and *BCDE*, for a total of four. This, together with an equal number of blocking coalitions in which he is critical, yields a Banzhaf index of 8.

Equivalent Voting Systems

[text pp. 431–435]

key idea

A **minimal winning coalition** is one in which each voter is critical to the passage of a measure; that is, if anyone defects, then the coalition is turned into a losing one.

question

Find the minimal winning coalitions in $[q : w(A), w(B), w(C)] = [7 : 5, 3, 2]$.

answer

$\{A, B\}$ and $\{A, C\}$.

explanation

The only other winning coalition is $\{A, B, C\}$, which is not minimal, since the coalition will win even if either B or C drops out.

key idea

Many voting systems are not presented as weighted voting systems, but are **equivalent** to weighted systems.

question

A small club has 5 members. A is the President, B is the Vice President, and C, D, and E are the ordinary members. The minimal winning coalitions are A and B, A and any two of the ordinary members, and B and all three of the ordinary members. Express this situation as an equivalent weighted voting system.

answer

There are many answers possible. One is {5 : 3, 2, 1, 1, 1].

explanation

Consider assigning a weight of 1 to each regular committee member. Then assign a weight to the Vice President and finally to the President with both based on the voting requirements. Another possibility would be to assign a weight of 2 to each regular committee member, producing: [9 : 5, 4, 2, 2, 2].

The Shapley–Shubik Power Index

[text pp. 435–441]

key idea

A **permutation** of voters is an ordering of all the voters. The first voter in a permutation whose vote would make the coalition a winning one is called the **pivotal voter** of that permutation. Each permutation has exactly one pivotal voter.

key idea

The **Shapley–Shubik power index** of a voter is the fraction of the permutations in which that voter is pivotal.

question

Calculate the Shapley–Shubik power index for each of the voters in the weighted voting system $[q : w(A), w(B), w(C), w(D)] = [8 : 6, 3, 2, 1]$.

answer

The answer is (2/3, 1/6, 1/6, 0).

explanation

A is pivotal in the following coalitions:

$$BAxx, CAxx, xxAx, \text{and } xxxA.$$

There are two permutations of the first and second cases, and 6 of the third and fourth cases, for a total of 16. B is pivotal in the following 4 coalitions:

$$ABCD, ABDC, ADBC, \text{and } DABC.$$

Similarly, C is pivotal in 4 coalitions:

$$ACBD, ACDB, ADCB, \text{and } DACB,$$

while D is a dummy. Hence the Shapley–Shubik indices are

$$(16/24, 4/24, 4/24, 0) = (2/3, 1/6, 1/6, 0)$$

key idea

When the number of voters is large, there are many permutations, and a direct calculation of the Shapley–Shubik index is difficult. If, however, many of the voters have equal votes, it is possible to compute this index by counting the numbers of permutations. A set of n members has $n! = n \times (n-1) \times (n-2) \times \cdots \times 2 \times 1$ permutations, which is called the **factorial** of n.

question

How many permutations are there of a set of 5 elements?

answer

The answer is $5! = 120$.

explanation

$5! = 5 \times 4 \times 3 \times 2 \times 1 = 120$.

question

Compute the Shapley–Shubik power index for the voting system $[q : w(A), w(B), w(C), w(D), w(E)] = [4 : 3, 1, 1, 1, 1]$.

answer

The answer is $(3/5, 1/10, 1/10, 1/10, 1/10)$.

explanation

A is pivotal in the following coalitions:

$$xAxxx, xxAxx, xxxAx.$$

There are 24 permutations of each of these three types, for a total of 72. For example, the four xs in $xAxxx$ constitute a permutation of the four remaining voters, B, C, D and E, and there are $4! = 24$ permutations of a set of four elements. Since there are a total of $5! = 120$ permutations, A's Shapley–Shubik index is $72/120 = 3/5$. The remaining four voters share equally the remaining $2/5$ of the power. Thus, each of them has an index $(2/5)/4 = 2/20 = 1/10$.

Systems with Large Numbers of Voters

[text pp. 441–443]

key idea

Analysis of voting systems with a large number of voters is possible only if all the voters, with only a few exceptions, are equally powerful.

PRACTICE QUIZ

1. What would be the quota for a voting system that has a total of 30 votes and uses a simple majority quota?
   a. 15
   b. 16
   c. 30

2. A small company has three stockholders: the president and vice president hold 6 shares each, and a long-time employee holds 2 shares. The company uses a simple majority voting system. Which statement is true?
   a. The long time employee is a dummy voter.
   b. The employee is not a dummy, but has less power than the officers.
   c. All three shareholders have equal power.

3. Which voters in the weighted voting system [12: 11, 5, 4, 2] have veto power?
   a. no one
   b. A only
   c. Both A and B

4. The weighted voting system [12: 11, 5, 4, 2] has how many winning coalitions?
   a. 1
   b. 3
   c. 8

5. If there are three voters in a weighted voting system, how many distinct coalitions of voters can be formed?
   a. 6
   b. 8
   c. 9

6. Given the weighted voting system [6: 4, 3, 2, 1], find the number of extra votes of the coalition {A, B, C}.
   a. 1
   b. 2
   c. 3

7. For the weighted voting system [10: 4, 4, 3, 2], which of the following is true?
   a. A has more power than C.
   b. A and C have equal power.
   c. C has more power than D.

8. What is the value of C^7_4?
   a. 210
   b. 35
   c. 28

9. Find the Banzhaf power index for voter B in the weighted voting system [12: 11, 5, 4, 3].
   a. 4
   b. 8
   c. 10

10. Calculate the Shapley-Shubik power index for voter B in the system [12: 11, 5, 4, 2].
    a. 1/12
    b. 3/12
    c. 8/12

CHAPTER 13

FAIR DIVISION

CHAPTER OBJECTIVES

Check off these skills once you feel that you have mastered them.

- ❑ Describe the goal of a fair-division problem.
- ❑ Define the term "player."
- ❑ Define the set-theoretic term "partition" and describe its application to a fair-division problem.
- ❑ List three different categories of fair-division problems.
- ❑ Explain what is meant by a continuous case fair-division problem and give an example.
- ❑ List two approaches to solving a continuous case fair-division problem.
- ❑ Explain what is meant by a discrete case fair-division problem and give an example.
- ❑ Describe a method for solving a discrete case fair-division problem.
- ❑ Calculate a discrete fair division for a small number of players and objects when:
  (a) each player has an equal share;
  (b) the players all have different shares.
- ❑ Use preference lists and the "bottom-up" approach to work out optimal strategies for two players taking turns to divide a collection of assets.

GUIDED READING

Introduction

[text pp. 454–455]

Fair-division problems arise in many situations, including divorce, inheritance, or the liquidation of a business. The problem is for the individuals involved, called the **players**, to devise a scheme for dividing an object or a set of objects in such a way that each of the players obtains a share that she considers fair. Such a scheme is called a **fair-division procedure**.

The Adjusted Winner Divorce Procedure

[text pp. 455–459]

key idea

In the **adjusted winner procedure** for two players, each of the players is given 100 points to distribute over the items that are to be divided. Each party is then initially given those items for which he or she placed more points than the other party.

question

Suppose that Joe and Mary place the following valuations on the three major assets which will be divided up:

Asset	Joe	Mary
House	40	60
Rolls-Royce	30	20
Beach House	30	20

Who gets each of the assets initially?

answer

Mary gets the house, while Joe gets the Rolls-Royce and the beach house.

explanation

The person who bids the most points on an item is initially assigned to that item.

question

How many points does each of the players get according to this allocation?

answer

60 points each.

question

Is any further exchange of property necessary in order to equalize the allocations?

answer

No, both Joe and Mary have received an equal number of points.

key idea

If the allocations of points differ, then an adjustment must be made.

question

Suppose that Joe and Mary change their evaluations, as follows:

Asset	Joe	Mary
House	30	60
Rolls-Royce	30	10
Beach House	40	30

Now, who gets each of the assets initially?

answer

Mary gets the house, while Joe gets the Rolls-Royce and the beach house.

question

How many points does each of the players get according to this allocation?

answer

Joe gets 70 points (30 for the Rolls-Royce and 40 for the beach house), while Mary gets 60 for the house.

question

What further exchange of property is needed to equalize the allocations?

answer

1/7 of the beach house is transferred to Mary.

explanation

Some of Joe's property has to be transferred to Mary, since he has received more points initially. To determine how much, we first compute the fractions

$$\frac{40}{30} \text{ (Beach house)} \qquad \frac{30}{10} \text{ (Rolls-Royce).}$$

We now transfer part of Joe's assets to Mary as follows: let x equal the fraction of the beach house which Joe will retain. Then, to equalize the number of points, we solve the equation

$$30 + 40x = 60 + 30(1 - x).$$

We find $x = 6/7$.

Since Joe retains 6/7 of the beach house, he must transfer $1 - 6/7 = 1/7$ of the beach house to Mary.

question

After this final reallocation of property, how many points does each of the players receive?

answer

64.3 points each.

explanation

Joe gets to keep the Rolls-Royce, which is worth 30 points to him, he also retains a 6/7 share in the beach house, which is worth $40(6/7) = 34.3$ points, giving him a total of 64.3 points. Mary gets the house, worth 60 points to her, and a 1/7 share in the beach house, worth $30(1/7) = 4.3$ points.

key idea

The adjusted winner procedure satisfies three important properties. The allocation must be:

(1) **equitable**
(2) **envy-free**
(3) **Pareto-optimal.**

question

Define each of the following terms: equitable, envy-free, and Pareto-optimal.

answer

Equitable: both players receive the same number of points.

Envy-free: neither player would be happier with what the other received.

Pareto-optimal: no other allocation can make one player better off without making the other player worse off.

The Knaster Inheritance Procedure

[text pp. 459–461]

key idea

The adjusted winner procedure applies only when there are two heirs. With three or more, the Knaster inheritance procedure can be used.

question

Joe, Mary, and Ted inherit a house. Their respective evaluations of the house are $105,000, $90,000, and $120,000. Describe a fair division.

answer

Ted gets the house, and pays $40,000 to Joe and $35,000 to Mary.

explanation

Since the highest bidder gets the asset, the house goes to Ted. Since Ted values the house at $120,000, his share is $40,000. Hence, he pays $80,000 into a kitty. Joe and Mary view their fare shares as $35,000 and $30,000, respectively. After they take these amounts from the kitty, $15,000 remains. This sum is then split equally among the three heirs.

question

Joe, Mary, and Ted are heirs to an estate, which consists of a house, an antique car, and a boat. Their evaluations of the items in the estate follow. Describe a fair division.

Objects	Joe	Mary	Ted
House	$120,000	$135,000	$100,000
Car	$40,000	$30,000	$22,000
Boat	$80,000	$60,000	$70,000

answer

Mary gets the house and pays $48,000 to Ted. Joe gets the antique car and the boat and pays $28,000 to Ted. Ted gets no item, but receives a total of $76,000 from Mary and Joe.

explanation

It is possible to solve this problem by doing a calculation similar to the previous one for each of the three items, and then adding up the totals. However, an alternative method is to look at the entire estate. For example, Joe's bids indicate that he places a total value of $240,000 on the estate, which means that his share is $80,000. Similarly, Mary's estimate is $225,000, with her share being $75,000, while Ted's estimate is $192,000, entitling him to $64,000. Now Joe gets the car and boat having a total value (in his estimate) of $120,000, which is $40,000 over his fair share. Hence, he places $40,000 into a kitty. In the same way, Mary gets the house and places $60,000 in the kitty. Ted then removes his share of $64,000, which leaves $36,000 in the kitty. This amount is then divided equally among the three heirs.

Taking Turns

[text pp. 461–464]

key idea

Two people often split a collection of assets between them using the simplest and most natural of fair-division schemes: **taking turns**. If they each know the other's preferences among the assets, their best strategies may not be to choose their own most highly preferred asset first.

example

Jim and Ed attend a car auction and together win the bidding for a collection of four classic cars: a '59 Chevy (C), a '48 Packard (P), a '61 Austin-Healey (A) and a '47 Hudson (H). They will take turns to split the cars between them, with Jim choosing first. Here are their preferences:

	Jim's Ranking	Ed's Ranking
Best	C	P
2nd best	P	H
3rd best	H	C
Worst	A	A

If Jim chooses the Chevy (his favorite) first, then Ed may choose his favorite, the Packard; Jim loses his second best choice. Is there a way for Jim to choose strategically to end up with both of his top two?

answer

Jim should choose the Packard first. Ed must now resort to choosing the Hudson (his second favorite). That leaves the Chevy available to Jim in the second round.

key idea

There is a general procedure for rational players, called the "bottom-up-strategy," for optimizing individual asset allocations when taking turns. It is based on two principles:

1. You should never choose your least-preferred available option.
2. You should never waste a choice on an option that will come to you automatically in a later round.

key terms and phrases

- taking turns

Divide-and-Choose

[text pp. 464–465]

key idea

A fair-division procedure known as divide-and-choose can be used if two people want to divide an object such as a cake or a piece of property. One of the people divides the object into two pieces, and the second person chooses either of the two pieces.

Cake-Division Schemes: Proportionality

[text pp. 466–472]

key idea

The divide-and-choose method cannot be used if there are more than two players. A **cake-division scheme** is a procedure that n players can use to divide a cake among themselves in a way which satisfies each player.

key idea

A cake-division scheme is said to be **proportional** if each player's strategy guarantees him a piece of size at least $1/n$ in his own estimation. It is **envy-free** if each player feels that no other player's piece is bigger than the one he has received.

key idea

When there are three players, the **lone-divider method** guarantees proportional shares.

question

Suppose that players 1, 2, and 3 view a cake as follows:

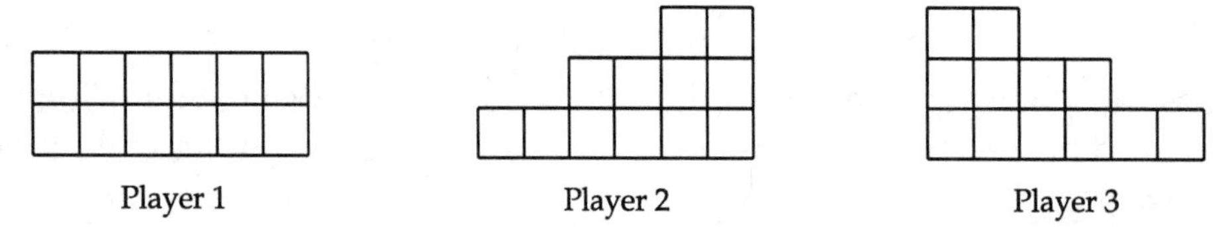

If player 1 cuts the cake into what she perceives as three equal pieces, draw three diagrams to show how each player will view the division.

answer

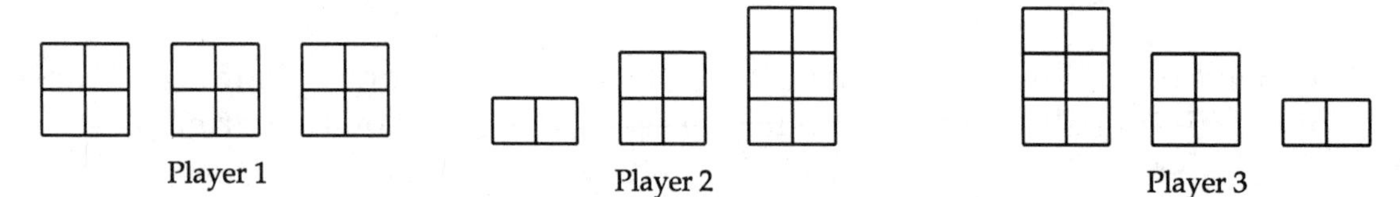

question

If player 2 is the first to choose, which piece will he pick, and how much of the cake does he believe he is getting?

answer

Player 2 will select the right-most piece, which he believes is 1/2 of the cake.

explanation

Player 2 believes that the pieces are cut unequally, with the right-most piece being the largest one.

question

Which of the two remaining pieces will player 3 choose, and how much of the cake does she believe she is getting?

answer

Player 3 will choose the left-most piece, which she believes is 1/2 of the cake.

explanation

Player 3 believes that the pieces are cut unequally, with the left-most piece being the largest.

question

Which piece will be left for player 1, and how much of the cake does she believe she is getting?

answer

The middle piece will remain for player 1, and she believes that it is 1/3 of the cake.

explanation

Player 1 believes that all three pieces are of equal size.

key idea

The **last-diminisher method** is a more complicated procedure which guarantees proportional shares with any number of players.

Cake-Division Schemes: The Problem of Envy

[text pp. 472–477]

key idea

Envy-free cake-division schemes are still more complicated. The **Selfridge-Conway procedure** solves this problem for the case of three players. For more than three players, a scheme which involves a **trimming procedure** has been developed. In it, proportions of the cake are successively allocated in an envy-free fashion, with the remaining portions diminishing in size. Eventually, the remainder of the cake is so small that it will not affect the perception of each player that he or she has obtained the largest piece.

If players cannot find a way to make an item fairly divisible, there may be no alternative but to sell it and share the proceeds equally.

PRACTICE QUIZ

1. In a fair-division procedure, the goal is for each person to receive
   a. an identical portion.
   b. what is perceived as an identical portion.
   c. what is perceived as an acceptable portion.

2. Using the adjusted winner procedure to divide property between two people, each person always receives
   a. exactly 50 points of value.
   b. at least 50 points of value.
   c. no more than 50 points of value.

3. Janna and George must make a fair division of three cars. They assign points to the cars, and use the adjusted winner procedure. Which car is divided between Janna and George?

object	Janna	George
Red car	40	20
White car	25	30
Blue car	35	50

a. Red car
b. White car
c. Blue car

4. Kim and Curt must make a fair division of three art items. They assign points to the items, and use the adjusted winner procedure. Who ends up with the Sculpture?

object	Kim	Curt
Painting	35	50
Sculpture	40	35
Tapestry	25	15

a. Kim
b. Curt
c. Kim and Curt share the Sculpture

5. Zac and Amy use the Knaster inheritance procedure to fairly divide a coin collection. Zac bids $700 and Amy bids $860. What is the outcome?
   a. Amy gets the coins and pays Zac $430.
   b. Amy gets the coins and pays Zac $390.
   c. Amy gets the coins and pays Zac $160.

6. Four children bid on two objects. Using the Knaster inheritance procedure, who ends up with the most cash money?

Object	Allison	Bruce	Charles	Darla
House	$81,000	$75,000	$82,000	$78,000
Car	$12,000	$11,000	$10,000	$13,000

a. Allison
b. Bruce
c. Darla

7. Four children bid on two objects. Using the Knaster inheritance procedure, what does Darla do?

Object	Allison	Bruce	Charles	Darla
House	$81,000	$75,000	$82,000	$78,000
Car	$12,000	$11,000	$10,000	$13,000

a. take car and take cash
b. take car and pay cash
c. take car only

8. Suppose Ben and Barb use the bottom-up strategy, taking turns to divide several items, ranked in order as shown below. If Ben goes first, what is his first choice?

Object	1st choice	2nd choice	3rd choice	4th choice
Ben	Clock	Radio	Toaster	Phone
Barb	Radio	Phone	Toaster	Clock

   a. Clock
   b. Radio
   c. Toaster

9. Suppose seven people will share a cake using the Last-diminisher method. To begin, Scott cuts a piece and passes it to Toni. Toni trims the piece and passes it on to each of the remaining five people, but no one else trims the piece. Then
   a. Scott gets this piece.
   b. Toni gets this piece.
   c. the last person who is handed the piece keeps it.

10. Which of the following procedures is envy-free?
   a. Lone-divider
   b. Selfridge-Conway
   c. Last-diminisher

CHAPTER 14

APPORTIONMENT

CHAPTER OBJECTIVES

Check off these skills once you feel that you have mastered them.

- ❑ State the apportionment problem.
- ❑ Explain the difference between quota and apportionment.
- ❑ State the quota condition and be able to tell which apportionment methods satisfy it and which do not.
- ❑ Do the same for the house monotone and population monotone conditions.
- ❑ Know that some methods have bias in favor of large or small states.
- ❑ Recognize the difference in computing quotas between the Hamilton method and divisor methods.
- ❑ Calculate the apportionment of seats in a representative body when the individual population sizes and number of seats are given, using the methods of Hamilton, Jefferson, Webster, and Hill-Huntington.
- ❑ Be able to give at least three reasons to support the claim that Webster's method is the "best" apportionment method.
- ❑ Calculate the critical divisor for each state.
- ❑ Determine an apportionment using the method of critical multipliers.
- ❑ Explain why the Jefferson method and the method of critical multipliers do not satisfy the quota condition.

GUIDED READING

Introduction

[text p. 489]

In many situations, a fixed number of places must be divided among several groups, in a way that is proportional to the size of each of the groups. A prime example of this is the division of the 435 seats in the U.S. House of Representatives among the 50 states. A problem arises because the exact allocations will usually involve fractional seats, which are not allowed. Various methods have been proposed to round the fraction to reach the total of 435, and four different methods have been used in the apportionment of the House during the past 200 years.

The Apportionment Problem

[text pp. 489–491]

key idea

The **apportionment problem** is to round a set of fractions so that their sum is a fixed number. An **apportionment method** is a systematic procedure that solves the apportionment problem.

key idea

The **population of the average congressional district** is obtained by dividing the total population by the house size. A state's **quota** is obtained by dividing its population by the average district population. It represents the exact share the state is entitled to. However, since the quota is usually not a whole number, an apportionment method must be used to change each of the fractions into an integer.

question

Suppose that a country consists of three states, *A*, *B*, and *C*, with populations 11,000, 17,500, and 21,500, respectively. If the congress in this country has 10 seats, find the population of the average congressional district and the quota of each of the three states.

answer

The average congressional district is 5000, and the quotas are *A* 2.2 seats, *B* 3.5 seats, *C* 4.3 seats.

explanation

The total population is 50,000 and there are 10 districts, so that the average size is 50,000 / 10 = 5000. We find each state's quota by dividing its population by 5000, the size of the average congressional district. Thus, *A*'s quota is 11,000 / 5000 = 2.2, *B*'s quota is 17,500 / 5000 = 3.5, and *C*'s is 21,500 / 5000 = 4.3.

question

States *A*, *B*, and *C* have populations of 11,000, 17,500, and 21,500, respectively. If their country's congress has 10 seats, which of the following assignments of seats fulfills the conditions of an apportionment method?

(a) *A* – 3 seats, *B* – 3 seats, *C* – 4 seats.
(b) *A* – 2 seats, *B* – 3 seats, *C* – 5 seats.
(c) *A* – 2 seats, *B* – 3 seats, *C* – 4 seats.
(d) *A* – 2 seats, *B*– 4 seats, *C* – 4 seats.

answer

(a), (b), and (d) are legitimate apportionments, while (c) is not.

explanation

In (c), the total number of seats allocated is 9, which is short of the house size of 10. The other three apportionments allocate 10 seats.

The Hamilton Method

[text pp. 492–496]

key idea

The **lower quota** for a state is obtained by rounding its quota *down* to the nearest integer, while to obtain the **upper quota** we round the quota *up*.

key idea

The **Hamilton method** first assigns each state its lower quota, and then distributes any remaining seats to the states having the largest fractional parts.

question

Consider the country previously listed, in which state *A* has a population of 11,000, state *B*, 17,500, and state *C*, 21,500. How would Hamilton allocate the 10 seats of the house?

answer

A will get two seats, while *B* and *C* each get four.

explanation

The quotas are: *A* – 2.2, *B* – 3.5, *C* – 4.3. Each state initially receives its lower quota, so that *A* receives two seats, *B* three, and *C* four. The one remaining seat now goes to the state with the largest fractional part, which is *B*. Hence, *B* also gets four seats.

question

Enrollments for four mathematics courses are as follows:

Algebra–62
Geometry–52
Trigonometry–38
Calculus–28

Ten mathematics sections will be scheduled. How many sections of each of these four courses will be allocated by the Hamilton method?

answer

There will be three sections of Algebra and Geometry, and two sections of Trigonometry and Calculus.

explanation

Since there are 180 students and 10 sections, the average size of a section will be 180 / 10 = 18. Thus, Algebra's quota is 62/18 = 3.44, while Geometry's quota is 52/18 = 2.89. Similarly, Trigonometry's quota is 2.11, and that of Calculus is 1.56. Distributing the lower quotas, Algebra gets three sections, Geometry and Trigonometry get two each, while Calculus gets one. The Hamilton method now assigns the remaining two sections to the courses having the largest fractional parts. Thus, Geometry gets the first section, and Calculus gets the second.

key idea

The **Alabama paradox** occurs if a state loses one or more seats when the number of seats in the house is increased. The **population paradox** occurs if a state loses at least one seat, even though its population increases, while another state gains at least one seat, even though its population decreases. The Hamilton method is susceptible to both of these paradoxes.

Divisor Methods

[text pp. 496–505]

key idea

A **divisor method** determines each state's apportionment by dividing its population by a common divisor *d* and rounding the resulting quotient. If the total number of seats allocated, using the chosen divisor, does not equal the house size, a larger or smaller divisor must be chosen. Finding a decisive divisor for a method of apportionment depends on how the fractions are rounded.

key idea

In the **Jefferson method** the fractions are all rounded *down.*

question

In the example of the three states (*A:* 11,000, *B:* 17,500, *C:* 21,500), would a divisor of 5000 be decisive in apportioning a house of 10 seats, according to the Jefferson method?

answer

The answer is no.

explanation

With a divisor of 5000, the quotas for *A, B,* and *C* are 2.2, 3.5 and 4.3, respectively. Since all fractions are rounded down in the Jefferson method, *A* would receive two seats, *B* three, and *C* four—for a total of nine, one fewer than the number to be allocated. Hence, 5000 is not a decisive divisor.

question

In the example of the three states (*A:* 11,000, *B:* 17,500, *C:* 21,500), would a divisor of 4350 be decisive for the Jefferson method?

answer

The answer is yes.

explanation

With a divisor of 4350, the quotas are: *A*–2.53, *B*–4.02, *C*–4.94. Rounding down gives *A* two seats, and *B* and *C* four each, for a total of ten.

key idea

Finding a decisive divisor by trial and error can be quite tedious. A systematic method for doing so involves calculating the **critical divisor** for each state.

question

Consider, again, the case of the four mathematics courses and their enrollment totals:

Algebra–62
Geometry–52
Trigonometry–38
Calculus–28

Ten mathematics sections will be scheduled. Use the method of critical divisors to determine how many sections of each of these courses will be allocated by the Jefferson method.

answer

Algebra gets four sections, Geometry gets three, Trigonometry gets two and Calculus gets one.

explanation

In the Jefferson method, fractions are rounded down. We begin with the quotas: [total population = 62 + 52 + 38 + 28 = 180; divisor = total population / # sections = 180/10 = 18; quota = population/divisor = 62/18 = 3.44, 52/18 = 2.89, 38/18 = 2.11, 28/18 = 1.56]. The tentative allocation of sections is three to Algebra, two each to Geometry and Trigonometry, and one to Calculus. To determine which courses get the two remaining sections, we compute the critical divisors for each course. The critical divisor for Algebra is 62/4 = 15.5. We obtain this by dividing Algebra's enrollment, 62, by 4, which is one more than its tentative allocation. Similarly, we compute 52/3 = 17.33, to obtain Geometry's critical divisor. Trigonometry's critical divisor is 12.67, and Calculus' is 14. Since Geometry's critical divisor is largest, it receives the next section. Before proceeding to allocate the next section, we recalculate Geometry's critical divisor, since it now has 3 sections. Its new divisor is 52/4 = 13. At this point, Algebra has the largest critical divisor, and it receives the last section.

key idea

The method of **critical multipliers** is a variation of the Jefferson method using multipliers instead of divisors. Each state starts with a **tentative apportionment** equal to its lower quota. For a state with tentative apportionment m and quota q, the critical multiplier is

$$\frac{n+1}{q}$$

The lowest of these multipliers is used to adjust the apportionments.

example

Suppose 100 students register for three science courses, as follows:

Biology – 43
Chemistry – 35
Physics – 22

If there is room to schedule only 7 sections all together, how many sections should be apportioned to each subject, according to the method of critical multipliers?

answer

The apportionment should be as follows: B gets 3 sections, C gets 3 sections, P gets 1 section.

☛ explanation

The quotas and tentative apportionments for each course are:

	quota	tentative apportionment
B	3.01	3
C	2.45	2
P	1.54	1
Total sections	6	

For example, the quota for C is obtained by multiplying the population (35) by the number of sections (7) and then dividing by the total population (100):

$$q = 35 \times 7 \;/\; 100 = 2.45$$

The tentative apportionment is the lower quota (2).
The critical multipliers can then be added to the table:

	quota	tentative app	crit mult
B	3.01	3	1.329
C	2.45	2	1.224
P	1.54	1	1.299

The lowest of the critical multipliers is that of C, 1.224. We use that to adjust the quotas, multiplying each quota by 1.224:

	quota	tentative app	crit mult	adjusted quota	adjusted apportionment
B	3.01	3	1.329	3.686	3
C	2.45	2	1.224	3	3
P	1.54	1	1.299	1.886	1
Total sections	7				

The classes are now fully apportioned.

✎ question

Suppose one student changes her mind and decides to switch from Chemistry to Physics. How does this affect the apportionment of sections to the three subjects?

✻ answer

This tiny change has a major effect; Chemistry loses an entire section, and Physics gains one section. The new apportionment is: B gets 3 sections, C gets 2 sections, P gets 12 sections.

☛ explanation

Here is the new table of quotas and adjustments:

	quota	tentative app	crit mult	adjusted quota	adjusted apportionment
B	3.01	3	1.329	3.739	3
C	2.38	2	1.261	2.957	2
P	1.61	1	1.242	2	2
Total sections	7				

You will notice that this one student caused the adjusted quota of Chemistry to fall just barely short of 3, with that of Physics just barely making it.

key idea

The Jefferson method and the method of critical multipliers do not satisfy the **quota condition:** an apportionment might not match either the upper or lower quota.

example

Suppose the enrollment figures in the above example were as follows:

Biology – 57
Chemistry – 22
Physics – 21

The quotas for the three subjects would be:

	quota	tentative apportionment
B	3.99	3
C	1.54	1
P	1.47	1
Total sections	5	

Applying the method of critical multiplers would lead, after two adjustment steps by two different multipliers, to this apportionment: B gets 5 sections, C gets 1 section, P gets 1 section. Thus, the apportioned number of sections for Biology (5) would exceed its upper quota of 4.

key terms and phrases

- tentative apportionment
- critical multiplier
- quota condition

key idea

The **Webster method** is also a divisor method, in which fractions greater than or equal to .5 are rounded up, while those less than .5 are rounded down.

question

Ten mathematics sections will be scheduled for four groups of students: Algebra (62), Geometry (52), Trigonometry (38), and Calculus (28). Use the method of critical divisors to determine how many sections of each of these courses will be allocated by the Webster method.

answer

Algebra and Geometry get three sections each, while Trigonometry and Calculus get two.

explanation

As we have seen, the quotas for the four subjects are 3.44, 2.89, 2.11, and 1.56, respectively. Now, however, we round the fractions in the normal way to obtain the Webster apportionment. Hence, Algebra, the fractional part of whose quota is less than .5, receives three seats in the tentative apportionment. Trigonometry's allocation is also rounded down to two, while Geometry and Calculus, both of whose fractional parts are greater than .5, have their allocations rounded up, Geometry to three and Calculus to two. Note: If the sum of the tentative allocations had been less than ten, we would have to find a smaller divisor. In either case, the method of critical divisors that we introduced in connection with the Jefferson method can be suitably modified to give the correct Webster apportionment. Trial and error can also be used to find an appropriate divisor.

key idea

In the Jefferson method, *all* fractions are rounded down, while in the Webster method, the cutoff point for rounding is .5. In the **Hill-Huntington method,** the cutoff point depends upon the size of the apportionment. If a state's quota is n seats, then its cutoff point is the geometric mean of n and $n + 1$, which is $\sqrt{n(n+1)}$. For example, if a state's quota is between 4 and 5, then $n = 4$, so that the cutoff point is $\sqrt{4 \cdot 5} = 4.472$. Thus, if a state's quota is 4.37, the state would get just 4 seats, since $4.37 < 4.472$. On the other hand, if its quota is 4.48, which is greater than the cutoff point of 4.472, it would get five seats.

question

Ten mathematics sections will be scheduled for four groups of students. Algebra (62), Geometry (52), Trigonometry (38), and Calculus (28). Use the method of critical divisors to determine how many sections of each of these courses will be allocated by the Hill-Huntington method.

answer

Algebra and Geometry each get three seats, while Trigonometry and Calculus get two each.

explanation

The quotas for the four courses are 3.44, 2.89, 2.11 and 1.56, respectively. We now compute the cutoff points for rounding up or down. For Algebra, the cutoff point is $\sqrt{3 \cdot 4} = \sqrt{12} = 3.46$. Since Algebra's quota is 3.44, which is less than the cutoff point, Algebra's tentative allocation is rounded *down.* The cutoff points for Geometry and Trigonometry are both $\sqrt{2 \cdot 3} = \sqrt{6} = 2.45$; so Geometry gets a third section, but Trigonometry does not. Finally, Calculus' cutoff point is $\sqrt{1 \cdot 2} = \sqrt{2} = 1.41$; so Calculus also gets an extra section, for a total of two. Since the sum of the tentative allocations is 10, the apportionment process is completed.

Which Divisor Method is the Best?

[text pp. 505–512]

key idea

All of the apportionment methods attempt to minimize inequities between the states, although they all use different criteria to measure the inequities.

key idea

The Webster method minimizes the inequity in the **absolute difference** in **representative shares.**

key idea

The Hill-Huntington method minimizes the relative difference in **representative shares** or **district populations.**

question

What is the relative difference between the numbers 8 and 13?

answer

The relative difference is (13 – 8) / 8 × 100% = .625 × 100% = 62.5%.

explanation

Because A > B, A = 13 and B = 8.

question

In the 1990 census, Alabama's population was 4,040,587, and Arizona's was 3,665,228. Alabama was apportioned 7 congressional seats, and Arizona received 6. What is the size of the average congressional district in each of these two states? Which of these two states is more favored by this apportionment?

answer

Alabama, whose average district population is 577,227, is favored over Arizona, whose average district population is 610,871. Each of these averages is found by dividing the state's population by the number of seats.

question

What is the relative difference in average district population between Alabama (577,227) and Arizona (610,871)?

answer

The relative difference is (610,871 – 577,227)/577,227 × 100% = .058 × 100% = 5.8%.

PRACTICE QUIZ

1. A county is divided into three districts with populations: Central, 3100; Western, 3500; Eastern, 1700. There are six seats on the county council to be apportioned. What is the quota for the Northeastern district?
   a. less than 1
   b. 1
   c. more than 1
2. The Hamilton method of apportionment can display
   a. the population paradox, but not the Alabama paradox.
   b. the Alabama paradox, but not the population paradox.
   c. both the Alabama paradox and the population paradox.
3. The Jefferson method of apportionment
   a. is a divisor method.
   b. satisfies the quota condition.
   c. is biased in favor of smaller states.
4. The Webster method of apportionment
   a. is susceptible to the Alabama paradox.
   b. favors smaller states.
   c. can have ties.

5. A county is divided into three districts with populations Central, 3100; Western, 3500; Eastern, 1700. There are nine seats on the school board to be apportioned. What is the apportionment for the Northeastern district using the Hamilton method?
   a. 1
   b. 2
   c. 3

6. A county is divided into three districts with populations: Central, 3100; Western, 3500; Eastern, 1700. There are nine seats on the school board to be apportioned. What is the apportionment for the Eastern district using the Jefferson method?
   a. 0
   b. 1
   c. 2

7. A county is divided into three districts with populations: Central, 3100; Western, 3500; Eastern, 1700. There are nine seats on the school board to be apportioned. What is the apportionment for the Eastern district using the Webster method?
   a. 1
   b. 2
   c. 3

8. A county is divided into three districts with populations: Central, 3100; Northern, 1900; Southern, 2800. There are five seats on the zoning board to be apportioned. What is the apportionment for the Southern district using the Hill-Huntington method?
   a. 1
   b. 2
   c. 3

9. The geometric mean of 7 and 8 is
   a. 7.5.
   b. more than 7.5.
   c. less than 7.5.

10. The relative difference between 7 and 8 is
    a. 1.
    b. 12.5%.
    c. 14.3%.

CHAPTER 15

GAME THEORY: THE MATHEMATICS OF COMPETITION

CHAPTER OBJECTIVES

Check off these skills once you feel that you have mastered them.

- ❑ Apply the minimax technique to a game matrix to determine if a saddlepoint exists.
- ❑ When a game matrix contains a saddlepoint, list the game's solution by indicating the pure strategies for both row and column players and the playoff.
- ❑ Interpret the rules of a zero-sum game by listing its payoffs as entries in a game matrix.
- ❑ From a zero-sum game matrix whose payoffs are listed for the row player, construct a corresponding game matrix whose payoffs are listed for the column player.
- ❑ If a two-dimensional game matrix has no saddlepoint, write a set of linear probability equations to produce the row player's mixed strategy.
- ❑ If a two-dimensional game matrix has no saddlepoint, write a set of linear probability equations to produce the column player's mixed strategy.
- ❑ When given either the row player or the column player's strategy probability, calculate the game's payoff.
- ❑ State in your own words the minimax theorem.
- ❑ Apply the principle of dominance to simplify the dimension of a game matrix.
- ❑ Construct a bimatrix model for an uncomplicated two-person game of partial conflict.
- ❑ Determine from a bimatrix when a pair of strategies is in equilibrium.
- ❑ Understand the role of sophisticated (vs. sincere) voting and the true power of a chair in small committee decision-making.
- ❑ Construct the game tree for a simple truel.
- ❑ Analyze the game tree of a truel, using backward induction to eliminate branches.

GUIDED READING

Introduction

[text pp. 525–526]

In competitive situations, parties in a conflict frequently have to make decisions which will influence the outcome of their competition. Often, the players are aware of the options—

called **strategies**—of their opponent(s), and this knowledge will influence their own choice of strategies. Game theory studies the **rational choice** of strategies, how the players select among their options to optimize the outcome. Some two-person games involve **total conflict,** in which what one player wins the other loses. However, there are also games of **partial conflict,** in which cooperation can often benefit the players.

Two-Person Total-Conflict Games: Pure Strategies

[text pp. 526–533]

key idea

The simplest games involve two players, each of whom has two strategies. The payoffs to each of the players is best described by a 2 × 2 **payoff matrix,** in which a positive entry represents a payoff from the column player to the row player, while a negative entry represents a payment from the row player to the column player.

question

Consider the payoff matrix

$$\begin{array}{c} \\ C \\ D \end{array} \begin{array}{c} \begin{array}{cc} A & B \end{array} \\ \begin{pmatrix} 3 & 4 \\ 2 & -5 \end{pmatrix} \end{array}$$

If the row player chooses C and the column player chooses *B*, what is the outcome of the game?

answer

The answer is 4.

explanation

The payoff associated with this outcome is the entry in Row C and Column *B*.

question

If the row player chooses C, what is the minimum payoff he can obtain? If he chooses *D*, what is his minimum payoff?

answer

The answer is 3; –5.

explanation

Recall that an outcome of –5 means that the row player loses 5 to the column player.

question

If the column player chooses *A*, what is the most she can lose? If she chooses *B*, what is the most she can lose?

answer

The answer is 3; 4.

explanation

An outcome of 3 means that the column player loses 3 to the row player.

key idea

We see in these examples that the row player can guarantee himself a payoff of at least 3 by playing *C*, and that the column player can guarantee that she will not lose more than three by playing *A*. The entry 3 is the minimum of its row, and it is larger than the minimum of the second row, –5. 3 is thus the **maximin,** and choosing *C* is the row player's **maximin strategy.** Similarly, 3 is the maximum of column *A*, and it is smaller than 4, which is the maximum of column *B*. Hence, 3 is the minimax of the columns, and if the column player chooses *A*, then she is playing her **minimax strategy.** When the maximin and minimax coincide, the resulting outcome is called a **saddlepoint.** The saddlepoint is the **value** of the game, because each player can guarantee at least this value by playing his/her maximin and minimax strategies. However, not every game has a saddlepoint. Games which do not will be studied in the next section.

question

Consider a game in which each of the players has a coin, and each chooses to put out either a head or a tail. (Note: The players do not flip the coins.) If the coins match, the row player wins, while if they do not match, the column player wins. The payoffs are as follows:

$$\begin{array}{c} \\ \textit{head} \\ \textit{tail} \end{array}\begin{array}{c} \begin{array}{cc} \textit{head} & \textit{tail} \end{array} \\ \begin{pmatrix} 2 & -4 \\ -3 & 5 \end{pmatrix} \end{array}$$

(a) What is the row player's maximin?
(b) What is the column player's minimax?
(c) Does this game have a saddlepoint?

answer

The answers are:
(a) –3
(b) 2
(c) no

explanation

If the maximin is different from the minimax, then there is no saddlepoint.

Two-Person Total-Conflict Games: Mixed Strategies

[text pp. 533–542]

key idea

When a game fails to have a saddlepoint, the players can benefit from using **mixed strategies,** rather than **pure strategies.**

key idea

The notion of **expected value** is necessary in order to calculate the proper mix of the players' strategies.

question

What is the expected value of a situation in which there are four payoffs, \$3, \$4, –\$2, and \$7, which occur with probabilities .2, .3, .45, and .05, respectively?

answer

The answer is \$1.25.

explanation

The expected value is found by multiplying each payoff by its corresponding probability and adding these products. We obtain

$$\$3 \times (.2) + \$4 \times (.3) - \$2 \times (.45) + \$7 \times (.05) = \$1.25.$$

question

Let's reconsider the game of matching coins, described by the payoff matrix

$$\begin{array}{c} \\ head \\ tail \end{array} \begin{array}{c} \begin{array}{cc} head & tail \end{array} \\ \begin{pmatrix} 2 & -4 \\ -3 & 5 \end{pmatrix} \end{array}$$

Suppose the row player mixes his strategy by choosing tails with probability p and heads with probability $1 - p$. If the column player always chooses heads, what is the row player's expected value?

answer

The answer is $2 - 5p$.

explanation

The expected value, $E_{head} = 2(1 - p) - 3p = 2 - 5p$.

question

$$\begin{array}{c} \\ head \\ tail \end{array} \begin{array}{c} \begin{array}{cc} head & tail \end{array} \\ \begin{pmatrix} 2 & -4 \\ -3 & 5 \end{pmatrix} \end{array}$$

Suppose the row player mixes his strategy by choosing tails with probability p and heads with probability $1 - p$. If the column player always chooses tails, what is the row player's expected value?

answer

The answer is $9p - 4$.

explanation

$E_{tail} = -4(1 - p) + 5p = 9p - 4$.

question

$$\begin{array}{c} \\ head \\ tail \end{array} \begin{array}{c} \begin{array}{cc} head & tail \end{array} \\ \begin{pmatrix} 2 & -4 \\ -3 & 5 \end{pmatrix} \end{array}$$

Using this diagram with its listed probabilities for the row players, find the best value of p, that is, the one which guarantees him the best possible return. What is the *(mixed-strategy) value* in this case?

answer

The answer is $p = 3/7$; value $= -1/7$.

explanation

The optimal value of p can be found in this case by setting E_{head} equal to E_{tail}, and solving for p. The equation is

$$9p - 4 = 2 - 5p,$$

or $14p = 6$, the solution to which is $p = 6/14 = 3/7$. The value of the game is $-1/7$, obtained by substituting this optimal value of p, which is 3/7, into either E_{head} or E_{tail}. This substitution yields $-1/7$.

question

Is this game **fair?**

answer

The answer is no.

explanation

Since the value of the game is negative, it is unfair to the row player.

key idea

A game in which the payoff to one player is the negative of the payoff to the other player is called a **zero-sum game.**

key idea

A zero-sum game can be **non-symmetrical** and yet fair.

example

Consider a coin-matching game with this payoff matrix:

$$\begin{array}{c c} & \begin{array}{cc} \textit{head} & \textit{tail} \end{array} \\ \begin{array}{c} \textit{head} \\ \textit{tail} \end{array} & \begin{pmatrix} 2 & 0 \\ -1 & -3 \end{pmatrix} \end{array}$$

This game is non-symmetrical and fair.

explanation

It is non-symmetrical because the payoffs for the row player are different from those for the column player. It is fair because the value of the game is 0; that payoff, when the row player chooses "head" and the column player chooses "tail," is a saddlepoint.

key idea

The **minimax theorem** guarantees that there is a unique game value, and an optimal strategy for each player. If this value is positive, then the row player can realize at least this value provided he plays his optimal strategy. Similarly, the column player can assure herself that she will not lose more than this value by playing her optimal strategy. If either one deviates from his or her optimal strategy, then the opponent may obtain a payoff greater than the guaranteed value.

Partial-Conflict Games

[text pp. 542–548]

key idea

In a game of total conflict, the sum of the payoffs of each outcome is 0, since one player's gain is the other's loss. **Variable-sum games,** on the other hand, are those in which the sum of the payoffs at the different outcomes varies. These are games of partial conflict, because, through cooperation, the players can often achieve outcomes that are more favorable than would be obtained by being pure adversaries.

key idea

In many games of partial conflict, it is difficult to assign precise numerical payoffs to the outcomes. However, the preferences of the parties for the various outcomes may be clear. In such a case, the payoffs are **ordinal,** with 4 representing the best outcome, 3 the second best, 2 next, and 1 worst. The payoff matrix now consists of pairs of numbers, the first number representing the row player's payoff, with the second number of the pair being the column player's payoff. Now, both like high numbers.

question

Consider the following matrix:

$$\begin{array}{cc} & \begin{array}{cc} A & B \end{array} \\ \begin{array}{c} C \\ D \end{array} & \begin{pmatrix} (1,3) & (2,2) \\ (4,1) & (3,4) \end{pmatrix} \end{array}$$

If the row player chooses *D* and the column player chooses *B*, what will the payoffs be to the players?

answer

The payoffs will be 3 to the row player and 4 to the column player.

explanation

The first entry in the outcome (3, 4) represents the payoff to the row player, and the second entry, the payoff to the column player.

question

Does either player have a **dominant strategy?**

answer

D is a dominant strategy for the row player. The column player does not have a dominant strategy.

explanation

The row player gets a better payoff in both cases by choosing strategy *D* (4 to 1 if the column player selects strategy *A*, and 3 to 2 if the column player selects strategy *B*).

The column player gets a more desirable payoff by switching from *A* to *B* when the row player selects strategy *C*; however, she gets a less desirable payoff by making the same switch when the row player selects strategy *D*.

key idea

When neither player can benefit by departing unilaterally from a strategy associated with an outcome, the outcome constitutes a **Nash equilibrium.**

question

$$\begin{array}{cc} & \begin{array}{cc} A & B \end{array} \\ \begin{array}{c} C \\ D \end{array} & \begin{pmatrix} (1,3) & (2,2) \\ (4,1) & (3,4) \end{pmatrix} \end{array}$$

If this outcome in the matrix is (1, 3), does either of the players benefit from defecting?

answer

The row player benefits by defecting to D, since he then obtains his best outcome (4), rather than his worst (1).

explanation

In the outcome (1, 2), the defection from C to D for the row player increases his payoff from 1 to 4. The defection from A to B for the column player, however, produces a payoff decrease.

question

$$\begin{array}{cc} & \begin{array}{cc} A & B \end{array} \\ \begin{array}{c} C \\ D \end{array} & \begin{pmatrix} (1,3) & (2,2) \\ (4,1) & (3,4) \end{pmatrix} \end{array}$$

Is there a Nash equilibrium in this matrix?

answer

(3, 4) is a Nash equilibrium.

explanation

Neither player can benefit by changing his or her strategy.

key idea

Prisoners' Dilemma is a game with the following matrix:

$$\begin{array}{cc} & \begin{array}{cc} A & D \end{array} \\ \begin{array}{c} A \\ D \end{array} & \begin{pmatrix} (2,2) & (4,1) \\ (1,4) & (3,3) \end{pmatrix} \end{array}$$

This matrix is also used to model other situations, such as an arms race. Here, A stands for "arm," and D for "disarm."

question

$$\begin{array}{cc} & \begin{array}{cc} A & D \end{array} \\ \begin{array}{c} A \\ D \end{array} & \begin{pmatrix} (2,2) & (4,1) \\ (1,4) & (3,3) \end{pmatrix} \end{array}$$

What is the most favorable outcome for the row player? For the column player?

answer

For the row player, it is the outcome where he selects A and the column player selects D. For the column player, it is the reverse.

☛. explanation

When the row player selects A and the column player D, the row player achieves his maximum payoff: 4.

When the column player selects A and the row player D, the column player achieves her maximum payoff: 4.

✎ question

$$\begin{array}{cc} & \begin{array}{cc} A & D \end{array} \\ \begin{array}{c} A \\ D \end{array} & \begin{pmatrix} (2,2) & (4,1) \\ (1,4) & (3,3) \end{pmatrix} \end{array}$$

Does the row player have a dominant strategy in Prisoner's Dilemma? How about the column player?

✲ answer

A is a dominant strategy for the row player and the column player.

☛ explanation

The row player always achieves a better payoff by selecting A rather than D (2 to 1 and 4 to 3). The column player fares similarly.

✎ question

$$\begin{array}{cc} & \begin{array}{cc} A & D \end{array} \\ \begin{array}{c} A \\ D \end{array} & \begin{pmatrix} (2,2) & (4,1) \\ (1,4) & (3,3) \end{pmatrix} \end{array}$$

Would it pay for either player to defect from the outcome (2, 2)? How about from the outcome (3, 3)?

✲ answer

No; yes.

☛ explanation

In the first case, the payoff to each defector would decrease from 2 to 1. In the second case, the payoff to each defector would increase from 3 to 4.

✎ question

Which outcome is a Nash equilibrium in Prisoners' Dilemma?

✲ answer

The answer is (2, 2).

☛ explanation

Neither player can benefit by defecting from the outcome (2, 2) because each reduces his or her payoff to 1.

key idea

Chicken is a game with the following payoff matrix:

$$\begin{array}{cc} & \begin{array}{cc} \textit{swerve} & \textit{don't swerve} \end{array} \\ \begin{array}{r} \textit{swerve} \\ \textit{don't swerve} \end{array} & \begin{pmatrix} (2,2) & (4,1) \\ (1,4) & (3,3) \end{pmatrix} \end{array}$$

question

$$\begin{array}{r} \\ \textit{swerve} \\ \textit{don't swerve} \end{array} \begin{array}{c} \begin{array}{cc} \textit{swerve} & \textit{don't swerve} \end{array} \\ \begin{pmatrix} (2, 2) & (4, 1) \\ (1, 4) & (3, 3) \end{pmatrix} \end{array}$$

What is the most favorable outcome for the row player? For the column player?

answer

For the row player it is when he doesn't swerve and the column player does. It is the reverse for the column player.

explanation

Each of these produces a maximum payoff: 4.

question

$$\begin{array}{r} \\ \textit{swerve} \\ \textit{don't swerve} \end{array} \begin{array}{c} \begin{array}{cc} \textit{swerve} & \textit{don't swerve} \end{array} \\ \begin{pmatrix} (2, 2) & (4, 1) \\ (1, 4) & (3, 3) \end{pmatrix} \end{array}$$

Does the row player have a dominant strategy? How about the column player?

answer

Neither player has a dominant strategy.

explanation

When the column player selects "swerve," the row player does better by selecting "not swerve;" however, the opposite is true for the row player's selection of "not swerve."

question

Would it pay for either player to defect from the outcome (3, 3)?

answer

It would pay for either player to defect, since he or she thereby obtains his or her most preferred outcome.

explanation

A detection for each player increased the payoff from 3 to 4.

question

Are there Nash equilibria in the game of Chicken?

answer

(2, 4) and (4, 2) are Nash equilibria.

explanation

Defecting from outcome (2, 4) decreases the row player's payoff from 2 to 1 and the column player's payoff from 4 to 3.

Defecting from outcome (4, 2) decreases the row player's payoff from 4 to 3 and the column player's payoff from 2 to 1.

Larger Games

[text pp. 549–561]

key idea

If one of three players has a dominant strategy in a 3 × 3 × 3 game, we assume this player will choose it and the game can then be reduced to a 3 × 3 game between the other two players. (If no player has a dominant strategy in a three-person game, it cannot be reduced to a two-person game.)

The 3 × 3 game is not one of total conflict, so the minimax theorem, guaranteeing players the value in a two-person zero-sum game, is not applicable. Even if the game were zero-sum, the fact that we assume the players can only rank outcomes, but not assign numerical values to them, prevents their calculating optimal mixed strategies in it.

The problem in finding a solution to the 3 × 3 game is not a lack of Nash equilibria. So the question becomes which, if any, are likely to be selected by the players. Is one more appealing than the others?

Yes, but it requires extending the idea of dominance to its successive application in different stages of play.

key idea

In a small group voting situation (such as a committee of three), **sophisticated voting** can lead to Nash equilibria with surprising results. An example is the **paradox of the chair's position**. This paradoxical weakness of the chair can be opposed through her use of a **tacit** or **revealed deception strategy**.

key idea

The analysis of a "truel" (three-person duel) is very different when the players move sequentially, rather than simultaneously.

example

Consider a sequential truel in which three perfect marksmen with one bullet each may fire at each other, each with the goal of remaining alive while eliminating the others. If the players act simultaneously, each has a 25% chance of survival. If they act sequentially, each will choose not to shoot, and all will survive.

explanation

If the players are A, B, C taking turns in that order, A cannot choose to shoot B (or C), because then C (or B) will shoot him next. A must pass. Similarly with B, and then C; none can risk taking a shot. Thus, all survive.

key idea

Sequential truels may be analyzed through the use of a **game tree**, examining it from the bottom up through **backward induction**.

key idea

The **theory of moves (TOM)** introduces a dynamic element into the analysis of game strategy. It is assumed that play begins in an initial state, from which the players, thinking ahead, may make subsequent moves and countermoves. Backward induction is the essential reasoning tool the players should use to find optimal strategies.

key terms and phrases

- non-symmetrical
- Prisoners' Dilemma
- Chicken
- sophisticated voting
- paradox of the chair's position
- tacit deception
- revealed deception
- game tree
- backward induction
- theory of moves (TOM)

Using Game Theory

[text pp. 561–565]

key idea

Game theory provides a framework for understanding the rationale behind conflict in our political and cultural world. An example is the confrontation over the budget between the Democratic President Bill Clinton and the Republican Congress that resulted in a shutdown of part of the federal government on two occasions, between November 1995 and January 1996. Government workers were frustrated in not being able to work, and citizens were hurt and inconvenienced by the shut down.

PRACTICE QUIZ

1. In the following two-person zero-sum game, the payoffs represent gains to Row Player I and losses to Column Player II.

$$\begin{bmatrix} 3 & 6 \\ 4 & 8 \end{bmatrix}$$

Which statement is true?
   a. The game has no saddlepoint.
   b. The game has a saddlepoint of value 4.
   c. The game has a saddlepoint of value 6.

2. In the following two-person zero-sum game, the payoffs represent gains to Row Player I and losses to Column Player II.

$$\begin{bmatrix} 4 & 7 & 1 \\ 3 & 9 & 5 \\ 8 & 2 & 6 \end{bmatrix}$$

What is the maximin strategy for Player I?
   a. Play the first row.
   b. Play the second row.
   c. Play the third row.

3. In the following two-person zero-sum game, the payoffs represent gains to Row Player I and losses to Column Player II.

$$\begin{bmatrix} 4 & 7 & 1 \\ 3 & 9 & 5 \\ 8 & 2 & 6 \end{bmatrix}$$

What is the minimax strategy for Player II?
   a. Play the first column.
   b. Play the second column.
   c. Play the third column.

4. In a two-person zero-sum game, suppose the first player chooses the second row as the maximin strategy, and the second player chooses the third column as the minimax strategy. Based on this information, which of the following statements is true?
   a. The game definitely has a saddlepoint.
   b. If the game has a saddlepoint, it must be in the second row.
   c. The game definitely does not have a saddlepoint.

5. In the game of matching pennies, Player I wins a penny if the coins match; Player II wins if the coins do not match. Given this information, it can be concluded that the two-by-two matrix which represents this game
   a. has all entries the same.
   b. has entries which sum to zero.
   c. has two 0s and two 1s.

6. In the following game of batter-versus-pitcher in baseball, the batter's batting averages are given in the game matrix.

		Pitcher	
		Fastball	Curveball
Batter	Fastball	.300	.200
	Curveball	.100	.400

What is the pitcher's optimal strategy?
   a. Throw more fastballs than curveballs.
   b. Throw more curveballs than fastballs.
   c. Throw equal proportions of fastballs and curveballs.

7. In the following game of batter-versus-pitcher in baseball, the batter's batting averages are given in the game matrix

		Pitcher	
		Fastball	Curveball
Batter	Fastball	.300	.200
	Curveball	.100	.400

What is the batter's optimal strategy?
   a. Anticipate more fastballs than curveballs.
   b. Anticipate more curveballs than fastballs.
   c. Anticipate equal proportions of fastballs and curveballs.

8. Consider the following partial-conflict game, played in a non-cooperative manner.

		Player II	
		Choice A	Choice B
Player I	Choice A	(3,3)	(4,1)
	Choice B	(1,4)	(2,2)

What outcomes constitute a Nash equilibrium?
   a. Only when both players select Choice A.
   b. Only when both players select Choice A or both select Choice B.
   c. Only when one player selects Choice A and the other selects Choice B.

9. True or False: Sequential and simultaneous trials result in different outcomes.
   a. True
   b. False

10. True or False: A deception strategy can help deal with the paradox of the chair's vote.
   a. True
   b. False

CHAPTER 16

GROWTH AND FORM

CHAPTER OBJECTIVES

Check off these skills once you feel you have mastered them.

- ❑ Determine the scaling factor when given the original dimensions of an object and its scaled dimensions.
- ❑ Given the original dimensions of an object and its scaling factor, determine its scaled dimensions.
- ❑ Calculate the change in area of a scaled object when its original area and the scaling factor are given.
- ❑ Calculate the change in volume of a scaled object when its original volume and the scaling factor are given.
- ❑ Determine whether two given geometric objects are similar.
- ❑ Locate on a number line the new location of a scaled point.
- ❑ When given the two-dimensional coordinates of a geometric object, its center, and the scaling factor, calculate its new coordinates after the scaling has taken place.
- ❑ Calculate from given formulas the perimeter and area of a two-dimensional object.
- ❑ Calculate from given formulas the surface area and volume of a three-dimensional object.
- ❑ Describe the concept of area-volume tension.
- ❑ Explain why objects in nature are restricted by a potential maximum size.

GUIDED READING

Introduction

[text pp. 581–582]

We examine how a variety of physical dimensions of an object—length, area, weight, and so on—are changed by proportional growth of the object. These changes influence the growth and development of an individual organism and the evolution of a species.

Geometric Similarity

[text pp. 582–584]

key idea

Two objects are **geometrically similar** if they have the same shape, regardless of their relative sizes. The **linear scaling factor** relating two geometrically similar objects A and B is the ratio of the length of a part of B to the length of the corresponding part of A.

question

Compare a cube A, which is 2 inches on a side, to a similar cube B, which is 12 inches on a side. What is the scaling factor of the enlargement from A to B and what is the ratio of the diagonal of B to the diagonal of A?

answer

The scaling factor is 6. The diagonal ratio is also 6.

explanation

The scaling factor r is the ratio of any two corresponding linear dimensions. Thus r = ratio of sides = 12/2 = 6, which is also the ratio of diagonals when comparing the second object to the original.

key idea

The area of the surface of a scaled object changes according to the square of the linear scaling factor.

question

Compare a cube A, which is 2 inches on a side, to a similar cube B, which is 12 inches on a side. What is the ratio of the surface area of the large cube B to that of the small cube A?

answer

The answer is 36.

explanation

You do not need to calculate the surface areas of the cubes. Area scales according to the square of the scaling factor. Since $r = 6$, $r^2 = 36$, and so we get (area B) / (area A) = 36.

key idea

The volume and weight of an object scale according to the cube of the linear scaling factor.

question

Cube A, which is 2 inches on a side, is similar to cube B, which is 12 inches on a side. If the small cube A weighs 3 ounces, how much does the large cube B weigh?

answer

The answer is 648 ounces, or 40.5 pounds.

explanation

Weight, like volume, scales by the cube of the scaling factor. Therefore, the ratio (weight B) / (weight A) = 6^3 = 216, and so weight B = 216 × 3 oz = 648 oz.

The Language of Growth, Enlargement, and Decrease

[text pp. 585–586]

key idea

In describing a growth situation or a comparison between two similar objects or numbers, the phrase "x is increased by" a certain percentage means that you must add the amount of the increase to the current value of x. The phrase "x is decreased by" a percentage means that you must subtract the amount of decrease from x.

question

This year, the value of the stock of the ABC Corporation rose by 25%, from 120 to ________.

answer

The answer is 150.

explanation

The new stock value is $120 + 25\% \times 120 = 120 + 30 = 150$.

question

This year, the value of the stock of the XYZ Corporation fell by 25%, from 150 to ________.

answer

The answer is 112.5.

explanation

The new stock value is $150 - 25\% \times 150 = 150 - 37.5 = 112.5$.

Numerical Similarity

[text pp. 586–588]

key idea

The Consumer Price Index (CPI) is the official measure of inflation. The CPI compares the current cost of certain goods, including food, housing, and transportation, with the cost of the same (or comparable) goods in a base period.

Measuring Length, Area, Volume, and Weight

[text pp. 588–592]

key idea

Some basic dimensional units in the US system of measurement are: foot (length), gallon (volume), pound (weight). Some comparable units in the metric system are: meter (length), liter (volume), kilogram (weight).

key idea

Conversions from one system to the other are done according to our rules for scaling. For example, 1 meter = 3.28 feet; that is, we have a scaling factor of 3.28 from meters to feet. Therefore, an area of $5\ m^2 = 5 \times (3.28)^2\ ft^2 = 53.79\ ft^2$.

question

Convert

5 pounds = __________ kilograms

60 kilometers = __________ miles

2000 in^2 = __________ m^2

answer

The answers are 5 lb = 2.27 kg; 60 km = 37.26 mi; 2000 in^2 = 1.29 m^2.

explanation

These answers are approximate:

1 lb = 0.4536 kg, so 5 lb = 5 × 0.4536 = 2.27 kg.

1 km = 0.621 mi, so 60 km = 60 × 0.621 = 37.26 mi.

1 in = 2.54 cm = 0.0254 m, so 1 in^2 = $(0.0254)^2$ = 0.000645 m^2; therefore, 2000 in^2 = 2000 × 0.000645 = 1.29 m^2.

Scaling Real Objects

[text pp. 592–595]

key idea

The size of a real object or organism is limited by a variety of structural or physiological considerations. For example, as an object is scaled upward in size, the **mass** and weight grow as the cube of the scaling factor, whereas the surface area grows as the square of the same factor.

key idea

The **pressure** on the bottom face or base of an object (for example, the feet of an animal or the foundation of a building) is the ratio of the weight of the object to the area of the base; that is, $P = W/A$. The weight may be calculated by multiplying the volume by the **density.**

question

What is the pressure at the base of a block of stone that is 2 ft wide, 3 ft long and 4 ft high, given that the density of the stone is 350 lb per ft^3?

answer

The answer is 1400 lb/ft^2.

explanation

$P = W/A$. The weight W of the block is volume × density. Thus $W = (2 \times 3 \times 4) \times 350$ lb/ft^3 = 8400 lb. The base area $A = 2 \times 3 = 6$ ft^2. Therefore, the pressure $P = 8400/6 = 1400$ lb/ft^2.

question

What are the dimensions and the pressure at the base of a block of the same stone that is the same shape but scaled up by a factor of 5?

answer

The answer is 7000 lb/ft^2.

explanation

The scaling factor is 5. Volume, and therefore weight, scale by $(5)^3 = 125$, while area scales by $(5)^2 = 25$. Thus, the pressure is $P = (125 \times W)/(25 \times A)$ $(125 \times 24 \times 350/25 \times 6) = 7000$ lb/ft^2.

question

What is the pressure at the base of a cylindrical column of the same stone that is 3 feet in diameter and 60 feet high, if the density of the stone is 350 lb/sq ft?

answer

The answer is 21,000 lb/ft^2.

explanation

The diameter of the column is 3, so the radius is 1.5. The column is a cylinder so its volume is $\pi \times (\text{radius})^2 \times \text{height} = 3.14 \times 2.25 \times 60 = 423.9$ ft^3. Because the density is 350 lb/ft^3, the weight W = volume × density = $423.9 \times 350 = 148{,}365$ lb. The area A of the base is $\pi \times (\text{radius})^2 = 7.065$ ft^2. Then $P = W/A = 148{,}365/7.065 = 21{,}000$ lb/ft^2.

Sorry, No King Kongs

[text pp. 595–599]

key idea

Though the weight of an object increases with the cube of the linear scaling factor, the ability to support weight increases only with the square of the linear scaling factor.

Solving the Problem of Scale

[text pp. 600–601]

key idea

As an object is scaled up in size, the area of its surface increases as the square of the scaling factor, while the volume increases as the cube of the scaling factor. This **area-volume tension** can strongly influence the development of structural parts of organisms that depend on both dimensions for strength, mobility, heat control, breathing, flight, and so on.

key idea

Proportional growth does not preserve many organic properties. As individuals grow, or as species evolve, their physiological proportions must change; growth is not proportional.

Falls, Jumps, and Flights

[text pp. 601–604]

key idea

Area-volume tension is a result of the fact that as an object is scaled up, the volume increases faster than the surface area and faster than areas of cross sections.

Keeping Cool (and Warm)

[text pp. 604–607]

key idea

Proportional growth is growth according to geometric similarity: the length of every part of the organism enlarges by the same linear scaling factor.

Proportional growth, scaling factor = 2

Disproportionate growth: height scaling factor = 1.6, length scaling factor = 2.86

Similarity and Growth

[text pp. 607–612]

key idea

Allometric growth is the growth of the length of one feature at a rate proportional to a power of the length of another.

Conclusion

[text p. 613]

key idea

Large changes in scale force a change in either material or form. Thus, limits are imposed on the scale of living organisms.

PRACTICE QUIZ

1. You want to enlarge a small painting, measuring 4 inches by 7 inches, onto a piece of paper, measuring 8.5 inches by 11 inches, so that the image is proportional and as large as possible. What is the scaling factor for the enlargement?
   a. 1.571
   b. 2.125
   c. 3.339

2. A small quilted wall hanging has an area of 15 sq. ft. A larger bed quilt is proportional to the wall hanging with a linear scaling factor of 3. What is the area of the bed quilt?
   a. 3 sq. ft.
   b. 45 sq. ft.
   c. 135 sq. ft.

3. A model airplane is built to a scale of 1 to 50. If the wingspan of the actual airplane is 25 ft, what is the wing span of the model plane?
   a. 2 ft
   b. 1/2 ft
   c. 1/10 ft

4. At the grocery store, an 8-inch cherry pie costs $4.39 and a similar 10-inch cherry pie costs $6.15. Which is the better buy?
   a. the 8-inch pie
   b. the 10-inch pie
   c. they are about the same

5. A 3rd grade class glues 64 sugar cubes together to form a larger cube. What is the linear scaling factor?
   a. 4
   b. 8
   c. 64

6. Your history book cost $43 in 1999. The earlier edition sold for $37 in 1992. Comparing the converted costs, which book is more expensive? (Assume the 1999 CPI is 166.2 and the 1992 CPI is 140.3.)
   a. the 1999 book
   b. the 1992 book
   c. The books are about the same price.

7. A home was purchased in 1976 for $25,000. How much would a comparable house be worth in 1999? (Assume the 1976 CPI is 56.9 and the 1999 CPI is 166.2.)
   a. about $25,290
   b. about $85,589
   c. about $73,932

8. A table weighs 125 pounds and is supported by four legs, which are each .75 inch by .5 inch by 24 inches high. How much pressure does each leg exert on the floor?
   a. 333 lb/sq. in
   b. 83 lb/sq in
   c. 2000 lb/sq in

9. The growth of human bodies is best modeled as ____________.
   a. isometric growth
   b. proportional growth
   c. allometric growth

10. If points (1, 5) and (2, 15) lie on the graph of $\log y = B + a \log x$, what is the value of a?
    a. (log 15 – log 5) / (log 2 – log 1)
    b. (log 15 – log 2) / (log 5 – log 1)
    c. (log 15 – log 1) / (log 5 – log 2)

CHAPTER 17

SYMMETRY AND PATTERNS

CHAPTER OBJECTIVES

Check off these skills once you feel you have mastered them.

- ❑ List the first ten terms of the Fibonacci sequence.
- ❑ Beginning with the number 3, form a ratio of each term in the Fibonacci sequence with its next consecutive term and simplify the ratio; then identify the number that these ratios approximate.
- ❑ List the numerical ratio for the golden section.
- ❑ Name and define the four transformations (rigid motions) in the plane.
- ❑ Analyze a given rosette pattern and determine whether it is dihedral or cyclic.
- ❑ Given a rosette pattern, determine which rotations preserve it.
- ❑ Analyze a given strip pattern by determining which transformations produced it.

GUIDED READING

Introduction

[text p. 629]

We study certain numerical and geometric patterns of growth and structure that can be used to model or describe an amazing variety of phenomena in mathematics and science, art and nature. The mathematical ideas the Fibonacci sequence leads to, such as the golden ratio, spirals, and selfsimilar curves, have long been appreciated for their charm and beauty; but no one can really explain why they are echoed so clearly in the world of art and nature. The properties of selfsimilarity, and reflective and rotational symmetry are ubiquitous in the natural world, and are at the core of our ideas of science and art.

Fibonacci Numbers

[text pp. 630–631]

key idea

Plants exhibiting phyllotaxis have a number of spiral forms coming from the sequence of **Fibonacci numbers.** This sequence {1, 1, 2, 3, 5, 8, 13, 21, 34, 55, 89, . . . } is generated according to the **recursion** formula that states that each term is the sum of the two terms preceding it.

question

Fill in the next five terms in the Fibonacci sequence {1, 1, 2, 3, 5, 8, 13, 21, 34, 55, 89, . . .}.

answer

The answer is: 144, 233, 377, 610, 987.

explanation

Using the recursion formula: $55 + 89 = 144$, $89 + 144 = 233$, $144 + 233 = 377$, $233 + 377 = 610$, $377 + 610 = 987$.

The Golden Ratio

[text pp. 631–637]

key idea

As you go further out in the sequence, the ratio of two consecutive Fibonacci numbers approaches the famous **golden ratio** $\phi = 1.618034\ldots$. For example, $89/55 = 1.61818\ldots$ and $377/233 = 1.618025\ldots$. The number ϕ is also known as the **golden mean.**

question

Calculate the square of ϕ, subtract 1, and compare the result to ϕ. What do you get?

answer

The answer is: ϕ.

explanation

ϕ satisfies the algebraic equation $\phi^2 = \phi + 1$, so $\phi^2 - 1 = \phi$.

question

Calculate the reciprocal $1/\phi$, add 1, and compare to ϕ. What do you get?

answer

The answer is: ϕ.

explanation

If you divide both sides of the equation $\phi^2 = \phi + 1$ by ϕ, you get $\phi = 1 + 1/\phi$

Balance in Symmetry

[text pp. 637]

key idea

Balance refers to regularity in how the repetitions are arranged. Along with similarity and repetition, balance is a key aspect of symmetry.

Rigid Motions

[text pp. 637–641]

key idea

Rigid motions are **translations, rotations, reflections,** and **glide reflections.**

question

Classify each of these rigid motions (within the given rectangle).

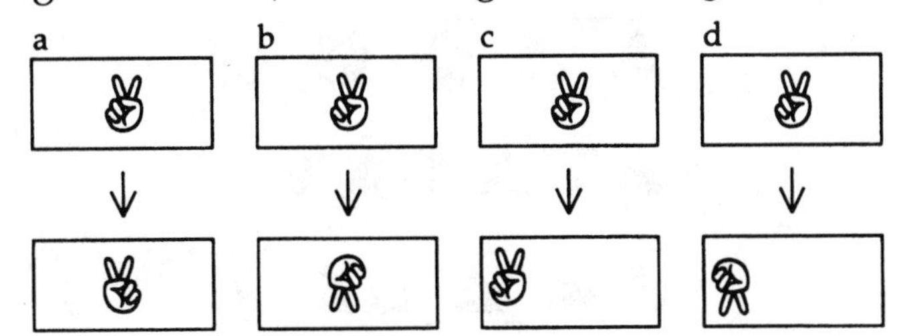

answer

The answer is:

a) reflection across a vertical line
b) rotation by 180°
c) translation to the left
d) glide reflection

explanation

a) this symmetry reverses left and right
b) this reverses left-right and up-down
c) the figure is moved to the left with the same orientation
d) the figure is moved to the left and simultaneously reversed up-down

Preserving the Pattern

[text pp. 641–643]

key idea

Preservation of the pattern occurs when the pattern looks exactly the same, with all parts appearing in the same places, after a particular motion is applied.

Analyzing Patterns

[text pp. 643–644]

key idea

Patterns are analyzed by determining which rigid motions preserve the pattern. These rigid motions are called symmetries of the pattern.

Strip Patterns

[text pp. 644–647]

key idea

Strip patterns repeat a design element along a line, so all of them have translation symmetry along the direction of the strip.

question

Which of the following linear designs are periodic strip patterns?

answer

The answer is pattern b.

explanation

In choice b, the design element consisting of two consecutive triangles is repeated at regular intervals. In choice a, there is no fixed repeated design element. In choice c, the intervals between repetitions are irregular.

key idea

Other possible symmetries for a strip pattern are horizontal or vertical reflection, rotation by 180°, or glide reflection.

key idea

Examples:

This strip pattern has vertical reflection, glide reflection, and rotation as symmetries, but not horizontal reflection.

♥♠♥♠♥♠♥♠

This strip pattern has horizontal reflection symmetry, but not vertical reflection or rotation.

››››››››

question

Determine all the symmetries of these strip patterns:

a) ❄❄❄❄❄❄❄❄ b) ↑↓↑↓↑↓↑↓

answer

The answer is:

a) all possible symmetries: translation, horizontal and vertical reflection, rotation, and glide reflection.
b) translation by two design units, vertical reflection, glide reflection, and rotation.

Symmetry Groups

[text pp. 647–651]

key idea

The full list of symmetries of any pattern forms a **symmetry group.** The combination of two symmetries A and B is written A • B.

question

If H = horizontal reflection and V = vertical reflection, what is $H \bullet V$?

answer

The answer is R_{180}.

explanation

H reverses up and down, V reverses left and right. R_{180} (a 180 degree rotation) reverses both, as does the combination symmetry $H \bullet V$. The effect can be traced visually as in this picture:

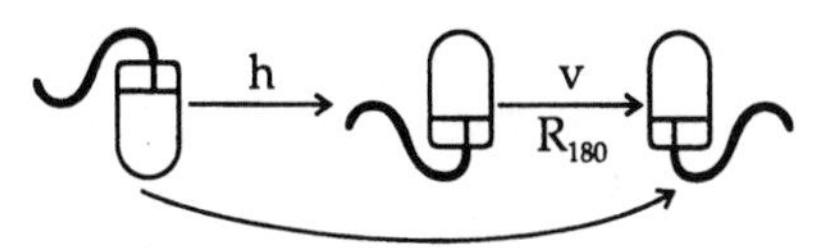

question

a) List the four elements of the symmetry group of a rectangle.
b) What is the combined result of applying all four of them consecutively?

answer

The answer is:

a) $\{I, H, V, R_{180}\}$
b) I

explanation

a) A rectangle is symmetrical across a vertical and horizontal axis, but not across the diagonals. It is, therefore, also symmetrical if rotated 180°, since $H \bullet V = R_{180}$.
b) To calculate the combination $I \bullet V \bullet H \bullet R_{180}$, first note $I \bullet V = V$. Then $V \bullet H = R_{180}$, and finally $R_{180} \bullet R_{180} = I$

key idea

Symmetry groups of rosette patterns contain only rotations and reflections.

question

What are the symmetries of these rosettes?

a) 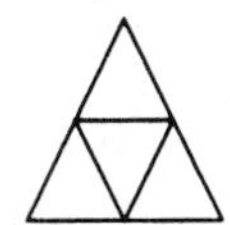

b) 

c) 

answer

The answer is:

a) Three rotations $\{I, R \bullet R^2$, where $R = R_{120}\}$ and three reflections across the axes a, b, v shown below:

b) Three rotations $\{I, R \bullet R^2$, where $R = R_{120}\}$ but no reflections, since the figure has a "right-handed" twist, and any reflection would change it to a "left-handed" one.
c) No symmetries other than I. The figure is totally asymmetric.

key idea

Symmetry groups of strip patterns include translation and may also contain glide reflections.

key idea

Glide reflection, symbolically G, is the combination of horizontal reflection with translation; that is, $H \cdot T = G$.

Notation for Patterns

[text pp. 651–652]

key idea

We can classify strip patterns according to their symmetry types; there are exactly seven different classes, designated by four symbols p * * * indicating the presence or absence of symmetry under translation, vertical reflection, horizontal or glide reflection, and rotation.

question

What is the symbolic notation for each of these patterns? You may need to refer to the flow chart on page 652 of the *For All Practical Purposes*, fifth edition, text to help answer this question.

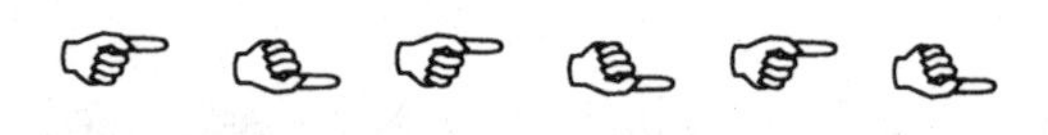

answer

The answer is:

a) p1a1
b) pmm2

explanation

a) No vertical or horizontal reflection, or rotation. However, each figure matches the next one over with a horizontal flip; that is, it has glide reflection symmetry.
b) This design has all possible symmetries for strip patterns.

key idea

It is useful to have a standard notation for patterns, for purposes of communication. Crystallographer's notation is the one most commonly used.

Imperfect Patterns

[text pp. 653–654]

key idea

In applying notation to patterns, it must be taken into account that patterns may not be perfectly rendered, especially if they are on a rounded surface.

PRACTICE QUIZ

1. The numbers 21 and 33 are consecutive numbers in the Fibonacci sequence. What is the next Fibonacci number after 33?
   a. 46
   b. 54
   c. 59

2. What is the geometric mean of 8 and 32?
   a. 20
   b. 16
   c. 6.32

3. The shorter side of a golden rectangle is 7 inches. How long is the longer side?
   a. 8.6 inches
   b. 9.5 inches
   c. 11.3 inches

4. Which figure has rotation symmetry?
   I. ⊖ II. ⊎
   a. I only
   b. II only
   c. Both I and II

5. Assume the following patterns continue in both directions. Which has a reflection isometry?
   I. TTTTTTTTTTTTT
   II. >>>>>>>
   a. I only
   b. II only
   c. Both I and II

6. Assume the following two patterns continue in both directions. Which of these patterns has a glide reflection isometry?
   I. <><><><><><><>
   II. ZZZZZZZZZZZZZ
   a. I only
   b. II only
   c. Neither I nor II

7. Use the flowchart in Figure 17.12 of *For All Practical Purposes* to identify the notation for the strip pattern below.

   ⇑⇓⇑⇓⇑⇓⇑⇓⇑⇓⇑⇓

   a. pma2
   b. p1a1
   c. p112

8. What isometries does this wallpaper pattern have?

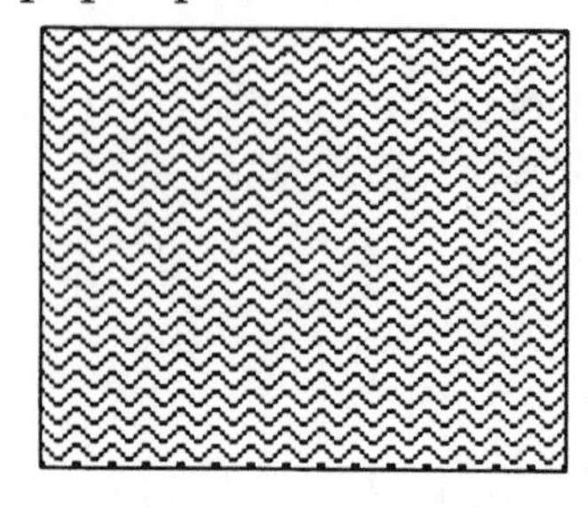

   a. translation only
   b. translation and reflection only
   c. translation, reflection, and rotation

9. What isometries does this wallpaper pattern have?

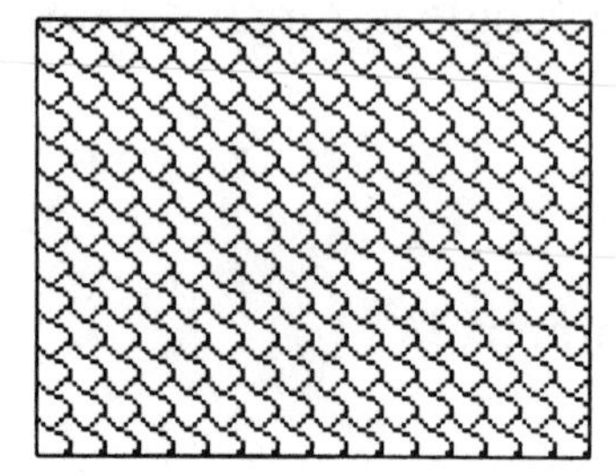

   a. translation only
   b. translation and reflection only
   c. translation, reflection, and rotation

10. How many elements are in the symmetry group of a regular hexagon?
    a. 6
    b. 12
    c. more than 12

CHAPTER 18

TILING

CHAPTER OBJECTIVES

Check off these skills once you feel you have mastered them

- ❑ Calculate the number of degrees in each angle of a given regular polygon.
- ❑ Given the number of degrees in each angle of a regular polygon, determine its number of sides.
- ❑ Define the term tiling (tessellation).
- ❑ List the three regular polygons for which a monohedral tiling exists.
- ❑ When given a mix of regular polygons, determine whether a tiling of these polygons could exist.
- ❑ Explain the difference between a periodic and a nonperiodic tiling.
- ❑ Discuss the importance of the Penrose tiles.
- ❑ Explain why fivefold symmetry in a crystal structure was thought to be impossible.

GUIDED READING

Introduction

[text pp. 668–669]

We examine some traditional and modern ideas about tessellation, the covering of an area or region of a surface with specified shapes. The beauty and complexity of such designs come from the interesting nature of the shapes themselves, the repetition of those shapes, and the symmetry or asymmetry of arrangement of the shapes.

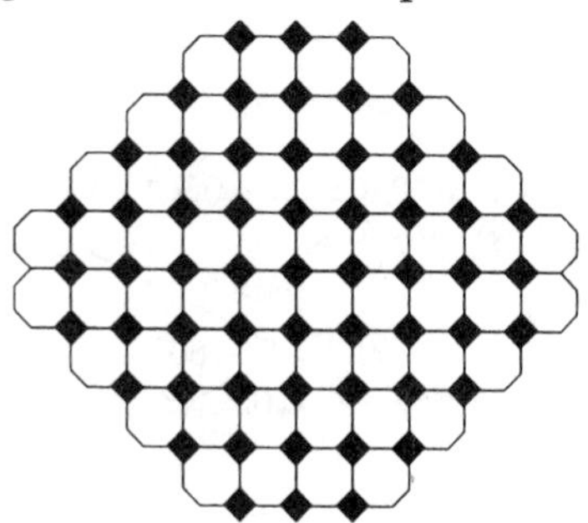

Regular Polygons

[text pp. 669–671]

key idea

A tiling is **monohedral** if all the tiles are the same shape and size; the tiling would consist of repetitions of one figure laid down next to each other.

question

Sketch an example of a monohedral tiling using a hexagon.

answer

The answer is: There are many kinds of hexagonal tilings; here is one example:

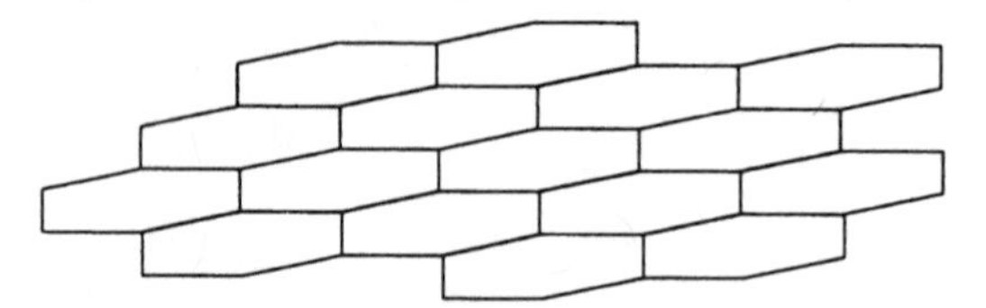

key idea

A tiling of the plane is a covering of that flat surface with non-overlapping figures.

question

a) Draw a sketch that shows how to tile the plane with squares of two different sizes.
b) Can you do the same thing with circles?

answer

The answer is:

a) The squares must fit together without spaces or overlaps. Here is a neat way to do it; there are many others.

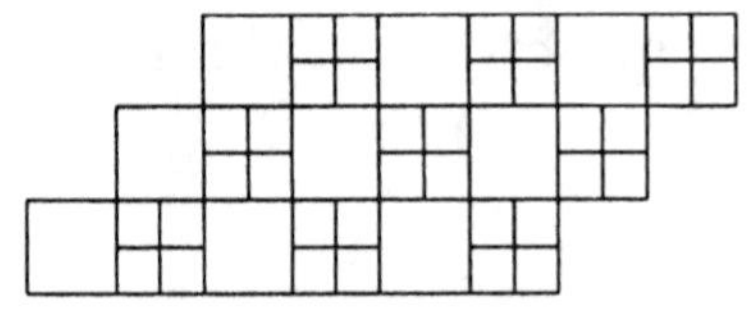

b) There is no way to fit circles together without overlapping or leaving spaces.

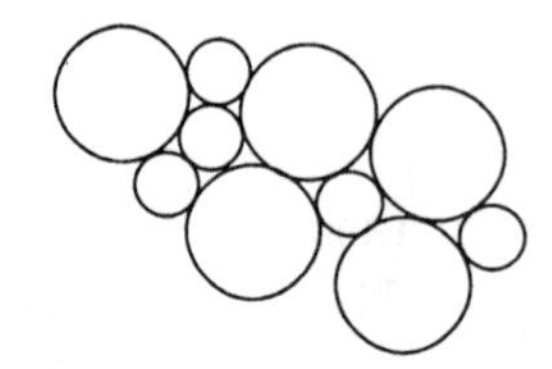

Regular Tilings

[text pp. 671–674]

key idea

In an **edge-to-edge tiling,** the interior angles at any vertex add up to 360°. For example,

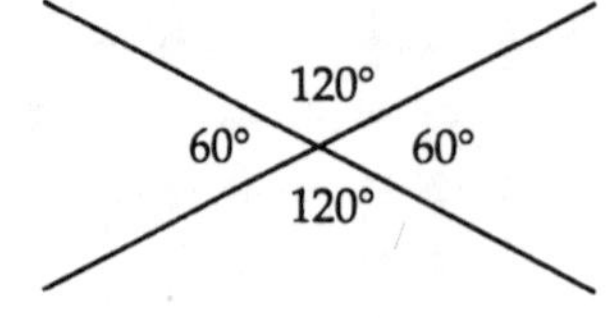

key idea

A **regular tiling** uses one tile which is a regular polygon. There are only three possible regular tilings:

Triangles

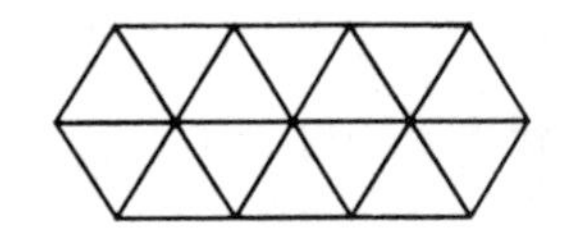

Hexagons

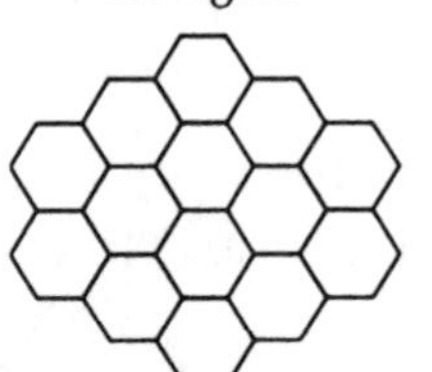

Squares

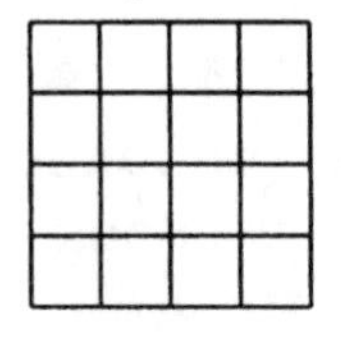

question

Explain why a regular edge-to-edge tiling using octagons is impossible.

answer

The answer is: A regular tiling using a polygon with n sides is impossible if $n > 6$. This is because the interior angle A, shown below for hexagons and octagons, would be larger than 120° (the angle for a hexagon), so three angles would not fit around a vertex, as it does in the hexagonal tiling.

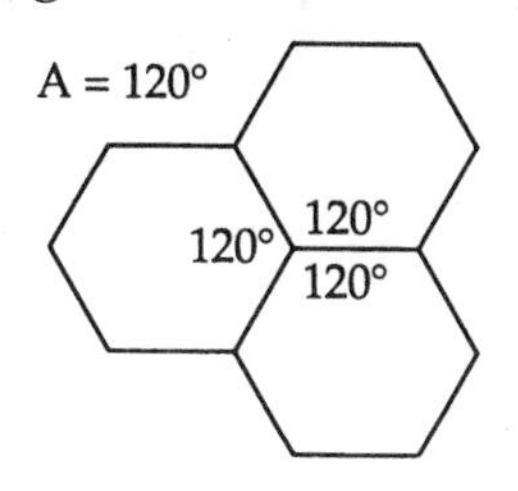

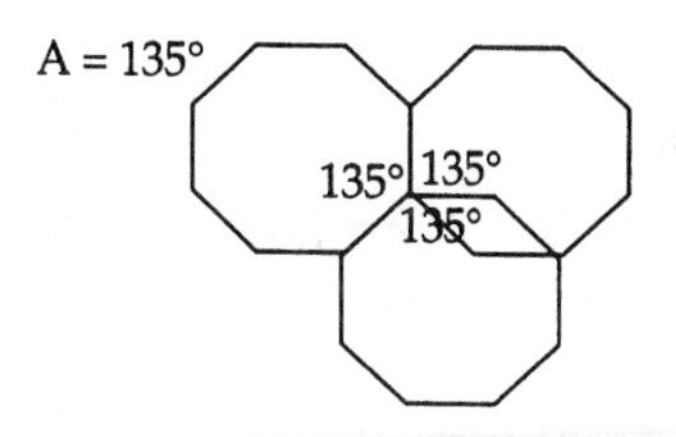

key idea

There are eight additional **semi-regular tilings,** using a mix of regular polygons with different numbers of sides.

Tilings with Irregular Polygons

[text pp. 675–680]

key idea

It is easy to adapt the square tiling into a monohedral tiling using a **parallelogram.** Since two triangles together to form a parallelogram, any triangle can tile the plane.

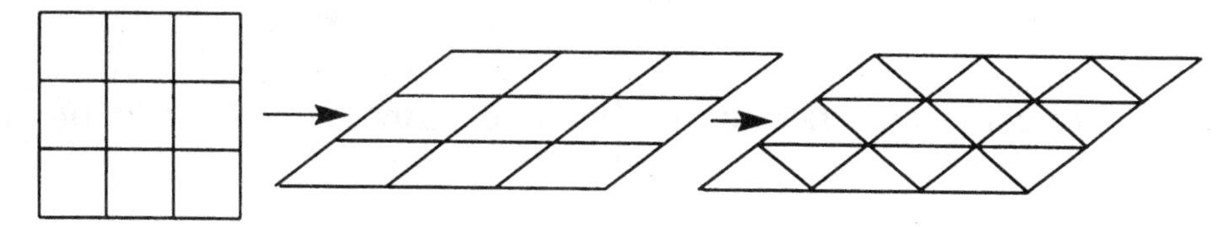

key idea

Any quadrilateral, even one that is not convex, can tile the plane.

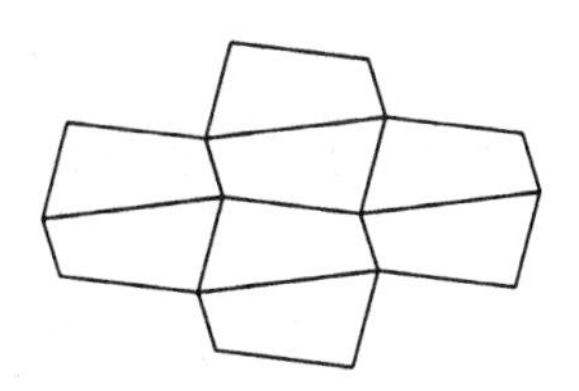

question

Draw a tiling of the plane with this figure:

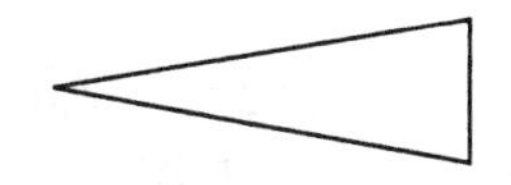

answer

Here is an edge-to-edge tiling with the given triangle, and another one which is not edge-to-edge:

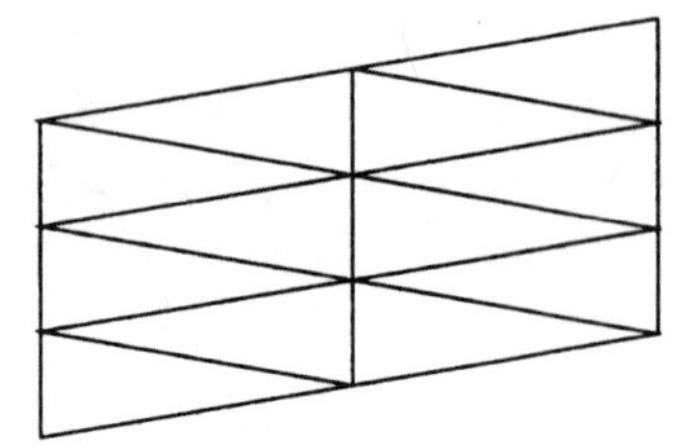

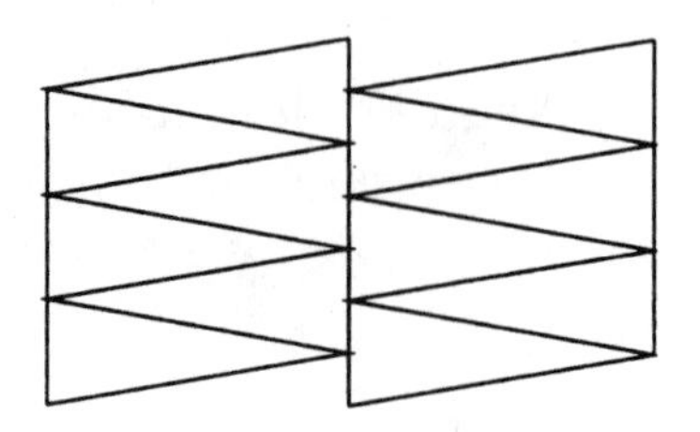

question

Draw a tiling of the plane with this figure:

answer

The answer is:

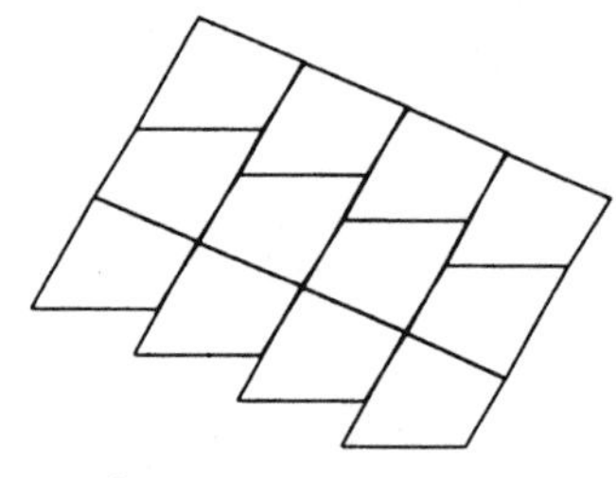

explanation

Two copies of the quadrilateral, one of which is rotated by 180°, fit together to form a parallelogram, thus forming an easy tiling.

key idea

Only certain classes of convex pentagons and hexagons can be used to tile the plane. A convex polygon with seven or more sides cannot tile the plane.

M. C. Escher and Tilings

[text p. 680]

key idea

The work of artist M. C. Escher, famous for his prints of interlocking animals, demonstrates an intimate link between art and mathematics.

Tiling by Translations

[text pp. 680–686]

key idea

The simplest way to create an Escher-like tiling is through the use of **translation.** The boundary of each tile must be divisible into matching pairs of opposing parts that interlock.

key idea

A single tile can be duplicated and used to tile by translation in two directions if certain opposite parts of the edge match each other.

Examples:

These two are based on a parallelogram tiling.

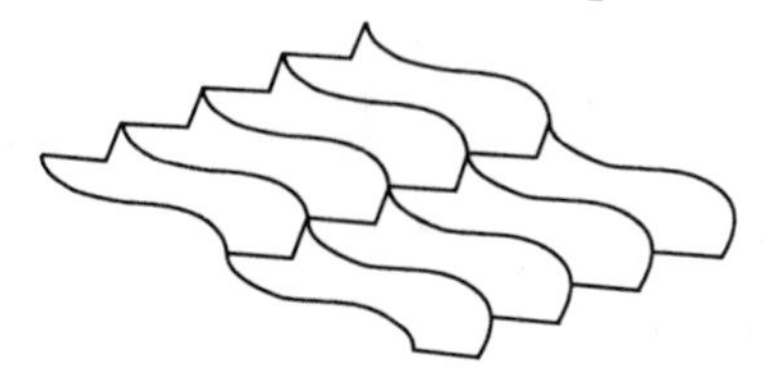

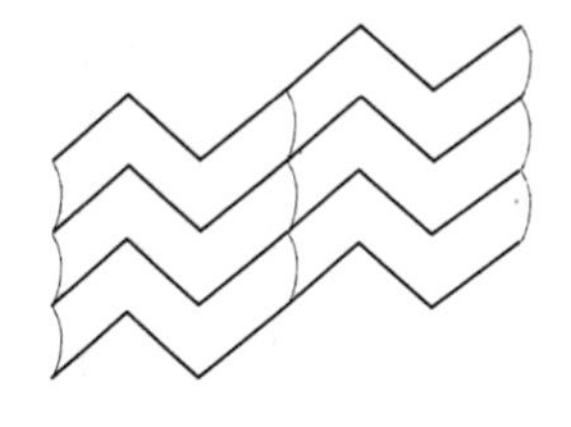

This one is based on a hexagon tiling.

question

a) Draw a square, replace opposite sides with congruent curved edges and draw a tiling by translation with the resulting figure.
b) Can you do the same thing with a triangle?

answer

The answer is:

a) Here is an illustration of the process for a square-based tile:

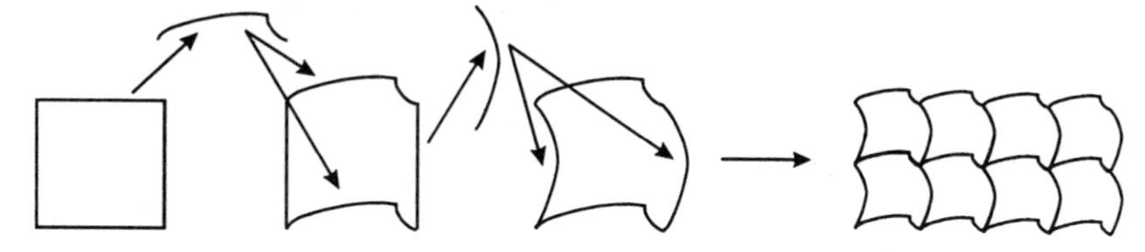

b) This cannot be done.

explanation

a) Translating horizontally and vertically, opposite sides match up.
b) You cannot pair up "opposite" sides in a triangle, or any polygon with an odd number of sides.

Tiling by Translations and Half-Turns

[text pp. 686–690]

key idea

If you replace certain sides of a polygon with matching **centrosymmetric** segments, it may be possible to use the resulting figure to tile the plane by translations and half-turns. The **Conway criterion** can be used to decide if it is possible.

✎ **question**

Start with this triangle:

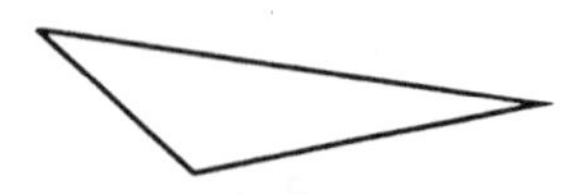

Replace the two long sides with centrosymmetric curves and sketch a tiling by translations and half-turns.

✲ **answer**

The answer is: This shows the process and the resulting tiling:

Here are the new sides

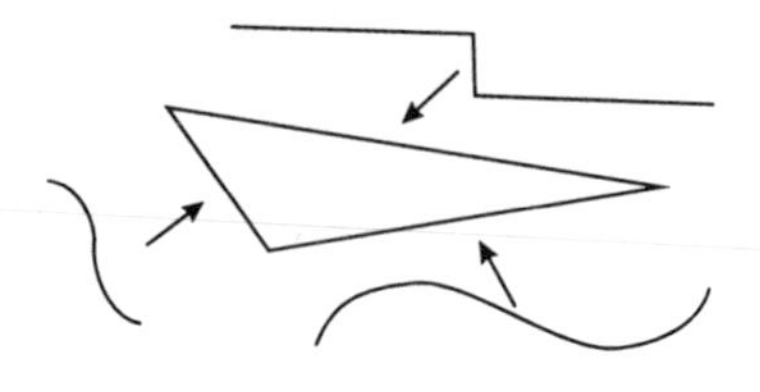

This is the tile

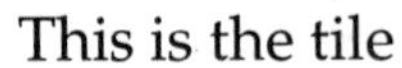

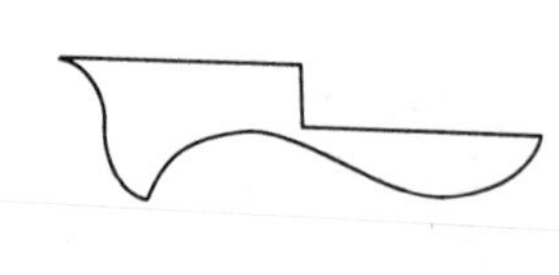

Here is the tiling:

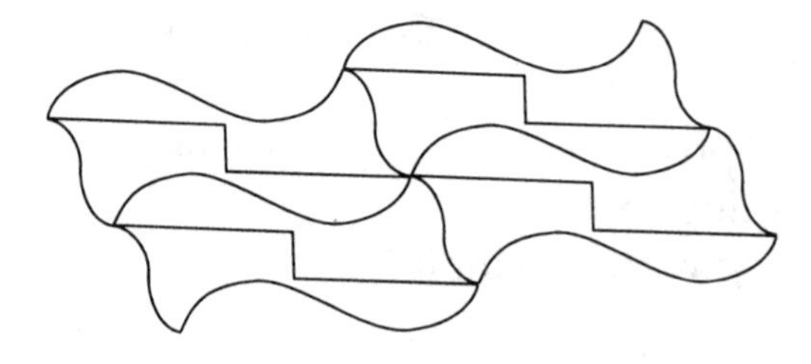

💡 **key idea**

Many fascinating and beautiful examples of these principles are found in the designs of the renowned graphic artist M. C. Escher.

Further Considerations

[text pp. 690–691]

💡 **key idea**

Periodic tilings have a **fundamental region** that is repeated by translation at regular intervals.

💡 **key idea**

A **fundamental region** consists of a tile, or block of tiles, with which you can tile a plane using translations at regular intervals.

✎ **question**

What is a fundamental region for this tiling?

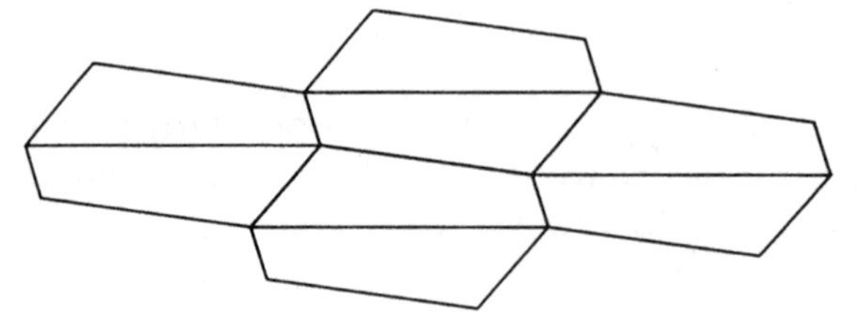

answer

The answer is:

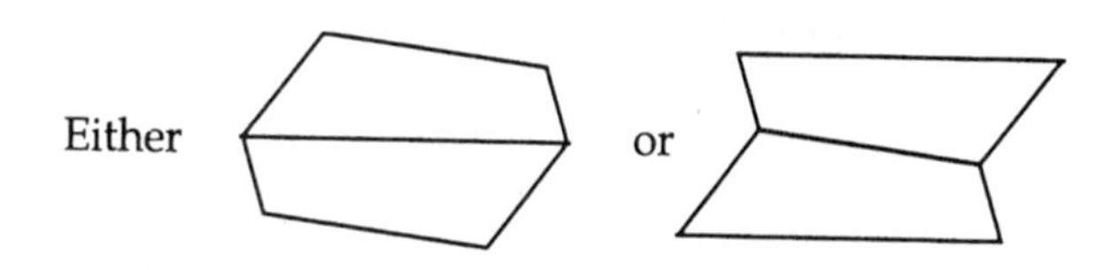

explanation

Two quadrilaterals which fit together form a hexagon that tiles by translation.

Nonperiodic Tilings

[text pp. 691]

key idea

A tiling may be **nonperiodic** because the shape of the tiles varies, or the repetition of the pattern by translation varies. The Penrose tiles are an important example of a set of two tiles which can be used only to tile the plane nonperiodically.

The Penrose Tiles

[text pp. 691–698]

key idea

Penrose tilings exhibit the following properties: self-similarity (inflation and deflation), the golden ratio (**quasiperiodic** repetition in that proportion), and partial five-fold rotational symmetry.

Quasicrystals and Barlow's Law

[text pp. 698–701]

key idea

Applying principles of tiling to three dimensional crystals, **Barlow's law** states that a crystal cannot have more than one center of fivefold rotational symmetry.

question

What chemical crystal exhibits strict fivefold symmetry?

answer

The answer is none; certain quasicrystals exhibit limited fivefold symmetry.

explanation

Crystals are periodic three-dimensional objects, and it follows from Barlow's law that truly periodic patterns can only have twofold, threefold, fourfold, or sixfold symmetry.

PRACTICE QUIZ

1. What is the measure of the exterior angle of a regular decagon?
    a. 36 degrees
    b. 72 degrees
    c. 144 degrees

2. Which of the following polygons will tile the plane?
    a. regular pentagon
    b. scalene triangle
    c. regular octagon

3. A semi-regular tiling is made with two dodecagons and another polygon at each vertex. Use measures of interior angles to determine which other polygon is required.
    a. triangle
    b. square
    c. hexagon

4. Squares and triangles can be used to form a semi-regular tiling of the plane. How many of each figure is needed?
    a. 2 triangles, 3 squares
    b. 4 triangles, 1 square
    c. 3 triangles, 2 squares

5. How many regular polyhedra exist?
    a. 3
    b. 6
    c. more than 6

6. Which of the following is true?
    I: Only convex polygons will tile the plane.
    II: Any quadrilateral will tile the plane.
    a. Only I is true.
    b. Only II is true.
    c. Both I and II are true.

7. It is possible to tile the plane with a square using
    I. Translations
    II. Half turns
    a. I only
    b. II only
    c. both I and II

8. Can the tile below be used to tile the plane?

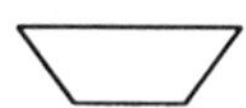

   a. Yes, using only translations
   b. Yes, using translations and half turns
   c. No

9. The Penrose tilings use two different figures with how many sides?
   a. 3 and 5
   b. 4 and 4
   c. 5 and 5

10. Which polygon can be altered to form an Escher-type tiling by translation only?
   a. equilateral triangle
   b. any convex quadrilateral
   c. regular hexagon

CHAPTER 19

LOGIC AND MODELING

CHAPTER OBJECTIVES

Check off these skills once you feel you have mastered them.

- ❑ Determine whether or not an English sentence is a proposition.
- ❑ Understand the meaning of the basic connectives, "not" and "implies" (symbolically $\neg$ and $\Rightarrow$).
- ❑ Form the contrapositive of an implication.
- ❑ Understand the meaning of a truth table for a proposition, and explain the truth tables for the basic connectives.
- ❑ Explain the relationships among the connectives "and," "or," "not" and "implies."
- ❑ Translate a sentence into a symbolic proposition, and vice versa.
- ❑ Use truth tables to determine if a proposition is satisfiable, or a tautology.
- ❑ Understand the role of rules of inference (such as modus ponens) in logical proofs.
- ❑ Use universal and existential quantifiers to form propositions in first-order logic.

GUIDED READING

Introduction

[text pp. 713–714]

Symbolic logic is the study of mathematical models of deductive reasoning. The most fundamental of these is **propositional logic**, a good model for analyzing our use of logical implication in everyday language. The next, more complex level is a model called **first-order logic**, well suited to understanding the deductive process in mathematics and science.

⇛ key terms and phrases:

- propositional logic
- first-order logic

The English Language

[text pp. 714–715]

key idea

The first goal of symbolic logic is to set up a formal language that captures the key aspects of a natural language like English, and which allows us to analyze the idea of "logical deduction."

The Basic Symbols of Propositional Logic

[text pp. 715–717]

key idea

Declarative sentences are "propositions"; we will model them using **sentence symbols** as notation. If A stands for a proposition, then the **negation** of A is written ¬A.

example

Let B stand for the sentence "Bob is hungry." The negation of B is:

¬B: "Bob is not hungry."

key idea

Implication is denoted by the "if-then" symbol ⇒. The proposition X ⇒ Y asserts that if X happens to be true, then Y must also be true. X is called the **antecedent**, and Y is called the **consequent** of the implication.

example

Let A stand for the proposition "It's after 5:00 pm." Then we can form this proposition:

A ⇒ B: If it's after 5:00 pm, then Bob is hungry

which expresses the idea that Bob inevitably gets hungry by 5, and stays hungry.

key terms and phrases:

- sentence symbols
- negation
- antecedent
- consequent

Using Symbolism in Analytical Reasoning

[text pp. 717–721]

key idea

The **contrapositive** of an implication is logically equivalent to the original implication; they are either both true or both false.

question

What is the contrapositive of the above statement A ⇒ B, "If it's after 5:00 pm, then Bob is hungry."

answer

We may informally state it this way: "If Bob is not hungry, then it can't be 5 pm yet."

explanation

The contrapositive of $A \Rightarrow B$ is $\neg B \Rightarrow \neg A$.

key idea

There are three common sense deductive rules which allow us to draw logical conclusions from implications.

example

One of the rules is this: If $X \Rightarrow Y$ is known, and Y is false, then X must also be false. For example, suppose what we said earlier about Bob is true: $A \Rightarrow B$ is known. If you run into Bob on the street, and he says he is not hungry, then you don't have to look at your wrist-watch, you can be sure it is earlier than 5 pm.

key terms and phrases:

- contrapositive

Truth and Falsity for ¬ and ⇒

[text pp. 721–723]

key idea

Propositions are declarative sentences with a definite **truth value**, which may be true or false, written T or F. A proposition P and its negation $\neg P$ have opposite truth values. This principle can be represented by the **truth table** for the symbolic sentence $\neg P$.

key idea

The truth table for the implication $P \Rightarrow Q$ is governed by the principle that $P \Rightarrow Q$ is only false in a situation in which P is true and Q is false.

question

Let's look at hungry Bob again. If it is 3:20 in the afternoon, and Bob is not hungry, what is the truth value of $A \Rightarrow B$?

answer

T

explanation

At 3:20 pm, the statement A is false; also, we know that B is false, since Bob says he is not hungry. So we are looking at this line in the truth table of $A \Rightarrow B$:

A	B	$A \Rightarrow B$
F	F	T

key terms and phrases:

- truth value
- truth table

The Other Connectives

[text pp. 724–729]

key idea

The connective "and" (written $\wedge$) is equivalent to a combination of $\neg$ and $\Rightarrow$, so it is not needed as a primitive symbol. This means that these two symbolic sentences

$$X \wedge Y \text{ and } \neg(X \Rightarrow \neg Y)$$

have the same truth table. In both cases, the sentence is true only if both X and Y are true; otherwise, it is false. Roughly speaking, a conjunction is the negation of an implication.

example

Let's write the proposition $X \wedge \neg Y$ using an implication.

a) $X \wedge \neg Y$ is equivalent to $\neg(X \Rightarrow \neg(\neg Y))$
b) We use the fact that $\neg(\neg Y)$ is equivalent to Y, one of the basic principles of our model.
c) We obtain $X \wedge \neg Y$ is equivalent to $\neg(X \Rightarrow Y)$

key idea

The connective "or" (written $\vee$) is also equivalent to a combination of $\neg$ and $\Rightarrow$. These two symbolic sentences

$$X \vee Y \text{ and } \neg X \Rightarrow Y$$

have the same truth table. In both cases, the sentence is true if either X is true or Y is true; it is false only when both X and Y are false. This is the inclusive, rather than the exclusive, use of the English word "or."

example

What is the truth table for the symbolic form of this sentence: "If I do not wash my car, it will not rain."

answer

Denote P: "I will wash my car"; and Q: "It will rain." Then the given proposition may be written symbolically as:

$$\neg P \Rightarrow \neg Q$$

question

Now write this sentence as an "or" statement.

answer

Either I will wash my car or it will rain.

explanation

The sentence $X \vee Y$ is equivalent to the or statement $\neg X \Rightarrow Y$.

Take P for X and $\neg Q$ for Y, and this says $P \vee \neg Q$ is equivalent to $\neg P \Rightarrow \neg Q$. This translates into the English sentence given in the answer.

key idea

A **tautology** is always true, for every possible assignment of truth values to its symbolic parts. A **satisfiable** sentence has some assignment of truth values which makes it true.

question

a) Is the sentence $P \Rightarrow (\neg P \vee Q)$ a tautology?
b) Is its negation satisfiable?

answer

a) No b) Yes

explanation

a) Here is the truth table, constructed step by step.

P	Q	$\neg P$	$\neg P \vee Q$	$P \Rightarrow (\neg P \vee Q)$
T	T	F	T	T
T	F	F	F	F
F	T	T	T	T
F	F	T	T	T

Thus, the sentence is false whenever P is true but Q is false.

b) The negation of the sentence, $\neg(P \Rightarrow (\neg P \vee Q))$, is true when T is true and F is false.

key idea

For very long sentences with a large number of symbolic parts, the **satisfiability problem** is very hard to solve. In general, the question of determining whether or not an arbitrary sentence is satisfiable is as hard as the Traveling Salesman Problem: they are both "NP-complete."

key terms and phrases

- tautology
- satisfiable
- satisfiability problem

A Model of Deduction

[text pp. 730–731]

key idea

Theorems are proven from axioms using rules of inference, such as **modus ponens**, **modus tollens**, and **double negation**. A demonstration of a theorem is called a **proof**.

key terms and phrases:

- modus ponens
- modus tollens
- double negation

First-Order Logic

[text pp. 732–734]

key idea

To model the language of ordinary mathematics, we need symbols for "quantifiers" that express existential ("there exists an x . . .") and universal ("for all x . . .") assertions.

example

Express symbolically this statement: "Everybody loves somebody."

explanation

We are making a statement about people, so we can represent an arbitrary person by letters x and y. The universal quantifier $\forall x \ldots$ asserts a fact about everybody, whereas the existential quantifier $\exists y \ldots$ asserts a fact about somebody. If we represent the property of person x "loving" person y as L(x,y), then we want to say that for every person x, there is a person y that x loves. Symbolically, that can be written:

$$\forall x \; \exists y \; L(x,y)$$

key idea

There is no algorithm for determining which statements in first-order logic are satisfiable or tautological.

PRACTICE QUIZ

1. Which implication is equivalent to "All freshmen are good students"?
   a. If s/he is a good student, then s/he is a freshman.
   b. If s/he is a freshman, then s/he is a good student.
   c. Neither a nor b

2. Using P for "I went to the party" and S for "I studied," translate into symbols the statement "I didn't study if I went to the party."
   a. $P \Rightarrow \neg S$
   b. $\neg S \Rightarrow P$
   c. $\neg(P \Rightarrow S)$

3. The implication $P \Rightarrow Q$ is false if
   a. P is false and Q is false.
   b. P is false and Q is true.
   c. P is true and Q is false.

4. The contrapositive of the statement "If it swims, then it isn't a cat" is
   a. If it is a cat, then it doesn't swim.
   b. If it doesn't swim, then it is a cat.
   c. If it is a cat, then it swims.

5. Which of the following is equivalent to $(P \vee Q) \Rightarrow (Q \wedge S)$?
   a. $[P \Rightarrow (Q \wedge S)] \vee [Q \Rightarrow (Q \wedge S)]$
   b. $(P \Rightarrow Q) \wedge (Q \Rightarrow R)$
   c. Neither a nor b

6. Build a truth table for $(P \vee Q) \Rightarrow (Q \wedge S)$. If P is true, Q is false, and R is true, then the statement is
   a. true.
   b. false.

7. The statement $(P \wedge Q) \Rightarrow Q$ is
   a. a tautology.
   b. satisfiable, but not a tautology.
   c. neither a nor b.

8. If a statement is not satisfiable, then its negation is a tautology.
   a. True
   b. False

9. Consider the following implications:
   If it is a dog, then it runs.
   Anything that has 4 legs, jumps.
   If it runs, then it jumps.

   Which of the following is a valid conclusion?
   I: If it is a dog, then it has 4 legs.
   II: If it has 4 legs, then it runs.
   a. I only
   b. II only
   c. Neither

10. Translate the following statement written in first order language:

$$(\forall x)(\exists y)\ [x < y]$$

   a. all x's are less than y's
   b. for all x's, there exists a larger y
   c. neither

CHAPTER 20

CONSUMER FINANCE MODELS

CHAPTER OBJECTIVES

Check off these skills once you feel you have mastered them.

- ❑ Apply the compound interest formula to calculate the balance of a savings account.
- ❑ Calculate the APY for a compound interest account.
- ❑ Describe the difference between arithmetic and geometric growth.
- ❑ List several applications of geometric growth.
- ❑ Use the savings formula to determine required deposits in a sinking fund, and payments required to fully amortize a loan.
- ❑ Calculate depreciation of a financial asset, given a negative growth rate.
- ❑ Determine the present value of currency in an inflationary context.
- ❑ Track the changing size of a population growing geometrically at a fixed rate.
- ❑ Understand the difference between the static reserve of a nonrenewable resource and its exponential reserve, and calculate the exponential reserve from the formula.
- ❑ Understand the meaning of a reproduction curve, and explain why a sustainable yield policy is needed for a harvestable resource.
- ❑ Estimate the maximum sustainable yield for a harvestable resource from its reproduction curve.

GUIDED READING

Introduction

[text p. 740]

We investigate and try to model situations of growth in economics and finance, examining the increase or decrease in value of investments and various economic assets. These models will apply similarly to the growth of biological populations, money in a bank account, pollution levels, and so on. In the same vein, managing a financial entity like a trust fund is similar to managing a renewable biological resource.

Models for Savings

[text pp. 740–742]

key idea

A savings account which earns compound interest is growing geometrically. At the end of the first year, the initial balance, or principal, is increased by the interest payment. Each successive year, the new balance is the previous balance plus interest, paid as a fixed percentage of that balance. The value of the savings account is determined by the **principal** (or **initial balance**), the rate of **interest** and the **compounding period**.

question

Suppose you deposit a $2000 principal to start up a bank account with an annual interest rate of 8%, compounded quarterly. How much money will you have in the account at the end of the first year?

answer

$2164.85

explanation

Each quarter the value of the account grows by 1/4 of the annual interest of 8%. This means the account balance is multiplied by 1.02 four times in the course of the year. Here is a table of account values:

initial deposit	quarter 1	quarter 2	quarter 3	quarter 4
$2000	$2040	$2080.80	$2122.41	$2164.85

key idea

The annual interest rate is called the **nominal rate** or the **APR**. With compounding, the actual realized percentage is higher; it is called the **effective rate**, **equivalent yield**, or the **APY**.

question

For the account above with APR of 8%, what is the APY?

answer

8.24%

explanation

The account realized an interest growth of $164.85 over the course of the year. As a percentage of the initial deposit, that is

$$165.85 / 2000 = .0824 = 8.24\%$$

key terms and phrases:

- principal
- initial balance
- interest
- compounding period
- nominal rate
- effective rate
- equivalent yield

The Mathematics of Geometric Growth

[text pp. 742–745]

key idea

The **compound interest formula** for the value of a savings account after compounding periods is

$$A = P(1 + r)^n$$

Here, P is the principal and r is the interest rate per compounding period.

question

If $1000 is deposited in an account earning 12% interest compounded annually, what will be the value of the account: a) after 5 years; b) after 20 years?

answer

a) $1762.34   b) $9646.29

explanation:

a) With annual compounding, we have one interest payment each year, or five in five years. Because r = 12% = .12, the formula for the value of the account after 5 years is $A = 1000(1.12)^5 = \$1762.34$.

b) After 20 years we have $A = 1000(1.12)^{20}$.

question

If $1000 is deposited in a different account earning 12% interest compounded monthly, what will be the value of the account: a) after 5 years; b) after 20 years?

answer

a) $1816.69   b) $10,892.55

explanation

a) With monthly compounding, we have 12 interest payments per year, or 60 in five years. Now the formula for the value of the account after 5 years is

$$A = 1000(1 + .12\ /\ 12)^{60} = 1000(1.01)^{60} = \$1816.69.$$

b) After 20 years, we have 240 interest periods, so $A = 1000(1.01)^{240} = \$10,892.55$.

key idea

If a population is experiencing **geometric (exponential) growth,** then it is increasing or decreasing by a fixed proportion of its current value with each measurement. The proportion is called the growth rate of the population.

key terms and phrases

- compound interest formula
- geometric growth
- exponential growth

Arithmetic Growth

[text pp. 745–747]

key idea

If a population, measured at regular time intervals, is experiencing **arithmetic growth** (also called **simple growth**), then it is gaining (or losing) a constant amount with each measurement. A financial account which is paying simple interest will grow arithmetically in value.

question

If an account has an initial value of $500 and gains $150 at each interval, write the sequence of population values for the first ten intervals.

answer

$500, 650, 800, 950, 1100, 1250, 1400, 1550, 1700, 1850

explanation

Starting with the initial value $500, successively add $150.

key terms and phrases:

- arithmetic growth
- simple growth

A Limit to Compounding

[text pp. 747–750]

Compounding more frequently at shorter intervals leads to a greater account value, because interest is earned earlier. However, this increase becomes progressively less significant as the compounding intervals get shorter. The limit is reached by continuous compounding, which is governed by the number $e \approx 2.71828$. The continuous interest formula for continuous compounding with interest rate r for m years is

$$A = Pe^{rm}$$

question

If $1000 is deposited in a continuous compounding account earning 12%, what will be the value of the account: a) after 5 years; b) after 20 years?

answer

a) $1822.11 b) $11,023.17

explanation

a) Using the calculator to evaluate $A = 1000\ e^{rm}$ with $r = .12$ and $m = 5$, we get $A = 1000\ e^{.6} =$ $1822.11.
b) Similarly, for $m = 20$, $A = 1000\ e^{2.4} =$ $11,023.17.

A Model for Accumulation

[text pp. 750–754]

key idea

We can accumulate a desired amount of money in a savings account by a fixed date by making regular deposits at regular intervals—a **sinking fund**. With a uniform deposit of d dollars at the end of each interval, and an interest rate of r per interval, the **savings formula** predicts that the value of the account after n intervals will be

$$A = (d / r)[(1 + r)^n - 1].$$

This formula is obtained by summing a **geometric series** of accumulated deposits and interest.

question

Rosetta's flower shop is growing, and she will need $6000 five years from now to pay for an addition to her greenhouse. How much would she have to deposit each month in a sinking fund with a 6% annual interest rate to accumulate this amount?

answer

Rosetta would deposit $86 at the end of each month for 5 years.

explanation

Here, d is the unknown amount to be deposited, r = 6% / 12 = .5% is the monthly interest rate and n = 60, the number of deposits in 5 years. Then we can solve for d in the savings formula:

$$6000 = (d / .005)\,[1.005^{60} - 1]$$
$$d = (6000)\,(.005) / .34885 = 85.996$$

key idea

When the regular amounts d are payments on a loan, they are said to **amortize** the loan. The **amortization formula** equates the accumulation in the savings formula with the accumulation in a savings account given by the compound interest formula. This is a model for saving money to pay off a loan all at once, at the end of the loan period.

key terms and phrases:

- sinking fund
- savings formula
- geometric series
- amortize
- amortization formula

A Model for Financial Derivatives

[text pp. 754–756]

key idea

An option to buy a stock at a certain price by a certain time is an example of a "financial derivative." The true value of a derivative depends on the current value, and the probabilities that the stock will go up or down within the option's time frame. The famous "Black-Scholes formula" is often used to value financial derivatives.

Exponential Decay

[text pp. 756–758]

key idea

Geometric growth with a negative growth rate is called exponential decay. Examples are depreciation of the value of a car, and decay in the level of radioactivity of a given quantity of a radioactive isotope. The quantity is declining at a rate which is negative and proportional to its size; the proportion λ is called the decay constant.

question

If a car costs $12,000 new and its value depreciates at 20% per year, give a formula for its value after n years, and predict its value in five years.

answer

The formula is $A = 12{,}000(.8)^n$ and the value will be $A = \$3932.16$.

explanation

The general formula is $A = P(1 + r)^n$, where $P = \$12{,}000$, $r = -.2$, $n = 5$. Use a calculator to evaluate $A = 12{,}000(.8)^5$.

key idea

With inflation, the value of currency declines. If the rate of inflation is i, then the **present value** of a dollar in one year is given by the formula

$$\text{present value} = 1 - i / (1 + i)$$

With successive years of inflation, this acts like a negative growth rate $r = -i / (1 + i)$.

question

Assuming constant 7% inflation, what would the present value of a $50 bill be in five years?

answer

$35.65

explanation

The present value of one dollar after one year is given by

$$pv(1) = 1 - .07 / 1.07 = .9346, \text{ or } 93.46 \text{ cents}$$

After five years, the present value has shrunk by this factor five times:

$$pv(5) = (1 - .07 / 1.07)^5 = .713, \text{ or } 71.3 \text{ cents}$$

Multiply that by 50 to get the answer.

key terms and phrases:

- present value

Growth Models for Biological Populations

[text pp. 758–760]

key idea

If we use a geometric growth model to describe and predict human (or other species) populations, the effective rate of growth is the difference between population increase caused by births and decrease caused by deaths. This difference is called the **rate of natural increase**.

example

To predict the population of the US in the year 2010, we can use the census information from 1990:

$$P = 255 \text{ million}, r = .7\% = .007.$$

With annual compounding, we can predict that the population 20 years later, in 2010:

$$P(2010) = 255 \text{ million} \times (1.007)^{02} = 293 \text{ million}.$$

At this rate, the population would double in about 100 years.

key idea

Small changes in the birth or death rate will affect the rate of natural increase, and this changes our prediction significantly.

question

Consider three countries A, B, and C, each of whose population in 1995 was 120 million. Country A is growing at a rate of .5%, country B at 1% and country C at 2%.

a) What would you predict the population of each country to be in 2010?
b) What would you predict for 2025?

answer

The table of population values for the three countries looks like this:

Country	1995	2010	2025
A	120 million	129 million	139 million
B	120 million	139 million	161 million
C	120 million	161 million	217 million

explanation

a) The year 2010 is 15 years later than 1995 so we use the formula

$$P(2010) = (120 \text{ million})(1 + r)^{15} \text{ with } r = .005 \text{ for A}, r = .01 \text{ for B, and } r = .02 \text{ for C}.$$

b) The year 2025 is 30 years after 1995 so $P(2025) = (120 \text{ million})(1 + r)^{30}$.

key terms and phrases:

- rate of natural increase

Nonrenewable Resources

[text pp. 760–763]

key idea

A **nonrenewable resource**, such as a fossil fuel or a mineral ore deposit, is a natural resource that does not replenish itself.

key idea

A growing population is likely to use a nonrenewable resource at an increasing rate. The regular and increasing withdrawals from the resource pool are analogous to regular deposits in a sinking fund with interest, and the same formula applies to calculate the accumulated amount of the resource that has been used, and is thus gone forever. The **static reserve** is the time the resource will last with constant use; the **exponential reserve** is the time it will last with use increasing geometrically with the population.

key idea

The formula for the exponential reserve of a resource with supply S, initial annual use U, and usage growth rate r is

$$n = \ln [1 + (S / U)r] / \ln [1 + r]$$

Here, ln is the natural logarithm function, available on your calculator.

question

Imagine that a certain iron ore deposit will last for 200 years at the current usage rate. How long would that same deposit last if usage increases at the rate of 4% each year?

answer

About 56 years.

explanation

The static reserve S / U is 200 years, so we can plug that value into the formula for n, using the assumed 4% = .04 for r. We obtain

$$n = \ln [1 + (200)(.04)] / \ln 1.04 = \ln 9 / \ln 1.04 = 2.197 / .0392 = 56 \text{ years}$$

key terms and phrases

- nonrenewable resource
- static reserve
- exponential reserve

Renewable Resources

[text pp. 763–772]

key idea

A **renewable natural resource** replenishes itself at a natural rate and can often be harvested at moderate levels for economic or social purposes without damaging its regrowth. Since heavy harvesting may overwhelm and destroy the population, economics and conservation are crucial ingredients in formulating proper harvesting policies.

key idea

We keep track of the population (measured in **biomass**) from one year to the next using a **reproduction curve**. Under normal conditions, natural reproduction will produce a geometrically growing population, but too high a population level is likely to lead to overcrowding and to strain the available resources, thus resulting in a population decrease.

example

This model leads to a reproduction curve looking something like this:

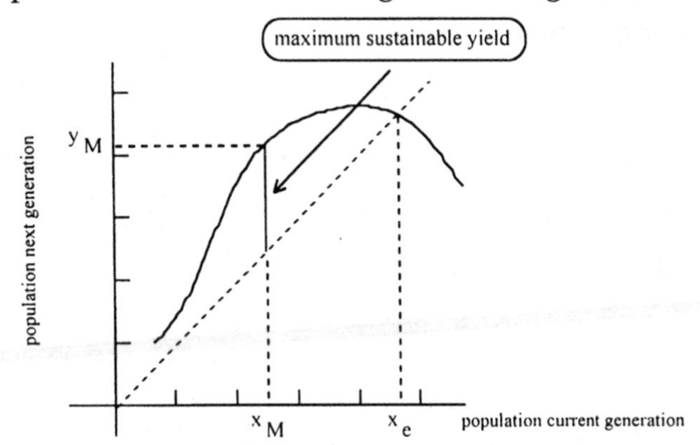

key idea

The dotted 45° diagonal line is the set of points where the population would be unchanged from year to year, and any point where it intersects the reproduction curve is an **equilibrium population size**.

key idea

The marked population value x_M is the level which produces maximum **natural increase** or **yield** in a year, and the difference between x_M and y_M (the population level a year later) is the **maximum sustainable yield** (or harvest) $h = y_M - x_M$. This amount is the maximum that may be harvested each year without damaging the population, and represents a good choice for a **sustained-yield harvesting policy**.

key idea

If our main concern is profit, we must take into account the economic value of our harvest, and the cost of harvesting. If we also include in our model **economy of scale** (denser populations are easier to harvest), then the sustainable harvest which yields a maximum profit may be smaller than the maximum sustainable yield and occurs at a population level $x_Q > x_M$.

key idea

Finally, if we also take into account the economic value of capital, and consider profit as our only motivation, it may be most profitable to harvest the entire population, effectively killing it, and invest the profits elsewhere. The history of the lumbering and fishing industries demonstrates this unfortunate fact.

key terms and phrases

- renewable natural resource
- biomass
- reproduction curve
- equilibrium population size
- natural increase
- yield
- maximum sustainable yield
- sustained-yield harvesting policy
- economy of scale

PRACTICE QUIZ

1. If you deposit $1000 at 5.2% simple interest, what is the balance after three years?
   a. $1140.61
   b. $1164.25
   c. $1156.00

2. Suppose you invest $250 in an account that pays 2.4% compounded quarterly. After 30 months, how much is in your account?
   a. $265.41
   b. $299.14
   c. $268.00

3. Suppose you deposit $10 at the end of each month into a savings account that pays 1.8% interest compounded monthly. After a year, how much is in the account?
   a. $121.60
   b. $130.41
   c. $131.96

4. What is the APY for 4.8% compounded monthly?
   a. 4.8%
   b. 4.9%
   c. 5.2%

5. If a bond matures in 3 years and will pay $5000 at that time, what is the fair value of it today, assuming the bond has an interest rate of 4.5% compounded annually?
   a. $4381.48
   b. $4325.00
   c. $4578.65

6. If you buy a home by taking a 30-year mortgage for $80,000, at an interest rate of 8% compounded monthly, how much will the monthly payments be?
   a. about $587
   b. about $805
   c. about $640

7. Suppose your stock portfolio increased in value from $12,564 to $12,870 during a one-month period. What is the APY for your gains during this period?
   a. 29.2%
   b. 32.6%
   c. 33.5%

8. Suppose you buy a new tractor for $70,000. It depreciates steadily at 8% per year. When will it be worth approximately $10,000?
   a. after about 11 years
   b. after about 18 years
   c. after about 25 years

9. Suppose a stock worth $50 today is equally likely to be worth $100 or $25 a year from today. What is a fair price to pay today for the option to buy this stock a year from now?
   a. about $7.50
   b. about $10
   c. about $12.50

10. Suppose a stock worth $50 today is equally likely to be worth $75 or $35 a year from today. What is the expected value of this stock a year from now?
    a. $75
    b. $55
    c. $50

ANSWERS TO POP QUIZZES

Chapter 1: 1b, 2b, 3b, 4c, 5a, 6b, 7a, 8a, 9b, 10c
Chapter 2: 1c, 2b, 3a, 4b, 5c, 6a, 7c, 8a, 9c, 10b
Chapter 3: 1c, 2c, 3a, 4b, 5b, 6c, 7c, 8a, 9c, 10c

Chapter 4: 1a, 2b, 3c, 4a, 5b, 6c, 7c, 8b, 9c, 10a
Chapter 5: 1b, 2c, 3a, 4a, 5a, 6a, 7b, 8b, 9b, 10a
Chapter 6: 1a, 2b, 3b, 4c, 5c, 6b, 7c, 8c, 9b, 10b

Chapter 7: 1c, 2b, 3a, 4b, 5b, 6c, 7a, 8a, 9b, 10c
Chapter 8: 1b, 2a, 3b, 4a, 5b, 6a, 7c, 8c, 9a, 10a
Chapter 9: 1a, 2a, 3a, 4c, 5c, 6c, 7a, 8a, 9a, 10a

Chapter 10: 1a, 2b, 3c, 4b, 5c, 6b, 7c, 8c, 9a, 10a
Chapter 11: 1a, 2c, 3c, 4a, 5b, 6a, 7c, 8b, 9c, 10c
Chapter 12: 1b, 2c, 3b, 4b, 5b, 6c, 7a, 8b, 9c, 10a

Chapter 13: 1c, 2b, 3b, 4c, 5b, 6a, 7a, 8b, 9b, 10b
Chapter 14: 1c, 2c, 3a, 4c, 5b, 6c, 7b, 8b, 9c, 10c
Chapter 15: 1b, 2b, 3c, 4b, 5b, 6c, 7a, 8a, 9a, 10a

Chapter 16: 1a, 2c, 3b, 4b, 5a, 6b, 7c, 8b, 9c, 10a
Chapter 17: 1b, 2b, 3c, 4a, 5c, 6a, 7a, 8c, 9b, 10b
Chapter 18: 1a, 2b, 3a, 4c, 5c, 6b, 7c, 8b, 9b, 10c

Chapter 19: 1b, 2a, 3c, 4a, 5a, 6b, 7a, 8a, 9c, 10b
Chapter 20: 1c, 2a, 3b, 4b, 5a, 6a, 7a, 8c, 9c, 10b